INTERNATIONAL WORKSHOP: BINARIES INSIDE AND OUTSIDE THE LOCAL INTERSTELLAR BUBBLE

INTERNATIONAL WORKSHOP: BINARIES INSIDE AND OUTSIDE THE LOCAL INTERSTELLAR BUBBLE

Santiago de Compostela, Spain 10 – 11 February 2011

EDITORS

J. A. Docobo
V. S. Tamazian
R. M. Aller Astronomical Observatory, Santiago de Compostela, Spain

Y. Y. Balega
Special Astrophysical Observatory of Russian Academy of Sciences, Karachai-Cherkessian Republic, Russia

SPONSORING ORGANIZATIONS
Ministry of Science and Innovation of Spain
University of Santiago de Compostela

AIP
American Institute of Physics

Melville, New York, 2012
AIP | CONFERENCE PROCEEDINGS ■ 1452

Editors

J. A. Docobo
V. S. Tamazian
R. M. Aller Astronomical Observatory
University of Santiago de Compostela
Avda. das Ciencias s/n.
Santiago de Compostela 15782
Spain

E-mail: joseangel.docobo@usc.es
vakhtang.tamazian@usc.es

Y. Y. Balega
Special Astrophysical Observatory
N. Arkhyz
Karachai-Cherkessian Republic 369167
Russia

E-mail: balega@sao.ru

L.C. Catalog Card No. 2012936510
ISBN 978-0-7354-1052-7
ISSN 0094-243X
Printed in the United States of America

AIP Conference Proceedings, Volume 1452
International Workshop: Binaries Inside and Outside the Local Interstellar Bubble

Table of Contents

PREFACE:

INTERNATIONAL WORKSHOP: BINARIES INSIDE AND OUTSIDE THE LOCAL INTERSTELLAR BUBBLE

The main lines of research of the Ramón María Aller Astronomical Observatory (OARMA) are the dynamics and astrophysics of double and multiple stars. R.M.Aller was the first person to introduce this class of studies in Spain at the beginning of the 20th century.

Since the renovation of OARMA during the past three decades, the Center has achieved a good scientific level not only regarding instrumentation (particularly with reference to the ICCD and EMCCD cameras for speckle interferometry and the 0.62m telescope) but also in terms of publications and service to the astronomical community with the edition of the Information Circular as well as the maintenance of the catalog of Double Star Orbits. Moreover, during the 2009-12 period, OARMA was the headquarters of the Presidency of the IAU Commission 26.

In 1996 and 2009, OARMA organized two international Workshops: "Visual Double Stars: Formation, Dynamics and Evolutionary Tracks", and "Double and Multiple stars: Dynamic, Physics, and Instrumentation". The present book contains the Proceedings of the third of this class of international scientific meetings. Our objective is always to highlight the efforts and accomplishments of the researchers who work in this area of Astronomy. All the topics addressed in this Workshop, e.g., physics, instrumentation and dynamics relative to double and multiple stars are within the framework of IAU Commission 26.

The Scientific Organizing Committee members were Drs Y.Y.Balega, J.A.Docobo (Chair), B.Mason, and V.S.Tamazian. The Local Organizing Committee was formed by Drs M.Andrade, P.P.Campo, J.A.Docobo (Chair), I.Fernandez, J.F.Ling, and V.S.Tamazian (Secretary).

The main sponsors were the Spanish Ministry of Science and Innovation and the University of Santiago de Compostela. We extend our sincere gratitude to the SOC and LOC, sponsors, and participants. The Proceedings volume includes the Invited Talks, Oral Contributions, and Posters presented at the meeting. We are indebted to the referees who carefully reviewed these papers.

J.A.DOCOBO, V.S.TAMAZIAN, and Y.Y.BALEGA

International Workshop: Binaries Inside and Outside the Local Interstellar Bubble
AIP Conf. Proc. 1452, 1-1 (2012); doi: 10.1063/1.4720579

Committees and sponsors

SCIENTIFIC ADVISORY COMMITTEE

- **Yuri Yu. Balega**
  Special Astrophysical Observatory
- **José A. Docobo** (Chairman)
  Astronomical Observatory R.M. Aller
- **Brian D. Mason**
  US Naval Observatory
- **Vakhtang S. Tamazian**
  Astronomical Observatory R.M. Aller

LOCAL ORGANIZING COMMITTEE

- **Manuel Andrade**
- **Pedro P. Campo**
- **José A. Docobo** (Chairman)
- **Iván Fernández**
- **Josefina F. Ling**
- **Vakhtang S. Tamazian** (Secretary)

SPONSORS

- **Spanish Ministry of Science and Innovation**
- **University of Santiago de Compostela**

Program

THURSDAY 10

- **Registration and inauguration**

10:00 Registration

10:15 Inauguration

- **Session I**

[Chair: Y. Balega]

10:30-11:20 Invited talk
H. Abt: *The age of the Local Interstellar Bubble.*

11:20-11:35 Coffee-break

11:35-11:50 Communication
J.A. Docobo, M. Andrade, P.P. Campo: *Guests in double stars.*

11:50-12:15 Break-poster session
J.A. Docobo, J.F. Ling, P.P. Campo: *Analyzing information contained in the OARMA Binary Orbit Catalog.*
M. Scardia, J.-L. Prieur, R. W. Argyle, L. Pansecchi, R. Gili, L. Koechlin: *Speckle interferometry with PISCO in Merate and prospects for the future.*

[Chair: N. Melikian]

12:15-13:05 Invited talk
B.D. Mason: *"Visual" orbit solutions from observing techniques - old and new.*

13:05-13:20 Communication
Y. Balega, E. Chentsov, V. Leushin, G. Weigelt: *Rotation velocities of θ_1 Ori C components and possible magnetic braking of the primary.*

- **Photo and lunch**

  13:45 Photo

  14:00 Lunch

- **Session II**

[Chair: B. Mason]

17:00-17:50 Invited talk
O. Malkov: *Eclipsing binaries: observational data, classification, and parametrization.*

17:50-18:10 Communication
J.A. Docobo, P.P. Campo: *The utilities of Docobo's method for the calculation of visual double star orbits using MATLAB.*

18:10-18:30 Coffe-break

18:30-19:20 Invited talk
N.D. Melikian, V.S. Tamazian, A.A. Karapetian, A.L. Samsoyan: *New emission objects and first flare stars in CYG OB7.*

- **Visit**

  19:45 Visit to OARMA

FRIDAY 11

- **Session III**

[Chair: H. Abt]

10:15-11:05 Invited talk
W.I. Hartkopf: *Plotting the future of the WDS.*

11:05-11:20 Coffee-break

11:20-11:30 Concluding remarks

- **Activities**

12:30	Visit to the Old Town of Santiago de Compostela
20:00	- Guitar Recital (Isabel Rei) - *Cantigas e Agarimos* Folk Group
21:30	Reunion dinner

The Age of the Local Interstellar Bubble

Helmut A. Abt

Kitt Peak National Observatory, abt@noao.edu

Keywords: Milky Way, Local Interstellar Bubble
PACS: 98.20.-d, 98.35.-a, 98.38.-j

The Local Interstellar Bubble is an irregular-shaped region that happens to be centered on the Sun; it has a minimum radius of 50 pc and a maximum radius of 150 pc. A map in the galactic plane and based on measures of the Na I D lines is shown in Figure 1. The interstellar density in the bubble is 200 times less than outside the bubble. The temperature is 10^6 K in the bubble compared with 8000 K outside. So it seems to be a region swept clear of most of the gas and heated by one or more supernovae. One wonders why the bubble has been stable for millions of years. The reason is that the outside density, ρ, is about 200 times greater than inside but the inside temperature, T, is roughly 200 times that outside, so ρT is balanced at the borders.

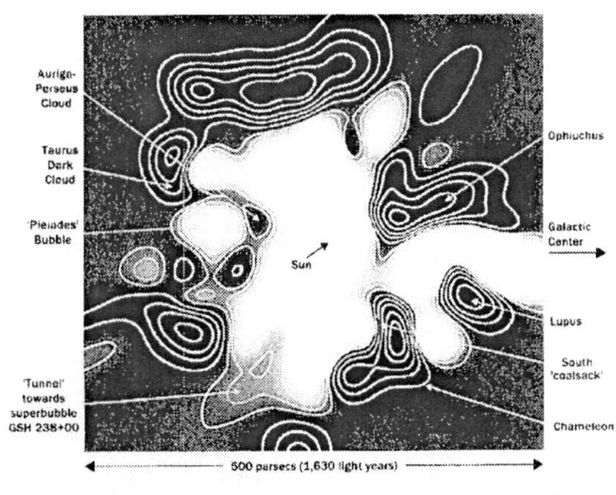

FIGURE 1. A map in the galactic plane of the Local Interstellar Bubble, adopted from measures of Na I D lines by Lalllement et al. 2003.

If the gas has been depleted by supernovae, then there would not be enough gas to form new stars. So I looked at the youngest stars in the bubble to tell the ages of the stars and therefore the times when the supernova or supernovae exploded. I used eight samples of stars, both singles and doubles. Let me discuss those eight samples in turn.

Figure 2 shows the distribution of absolute magnitudes of the primaries of 486 visual binaries (VB) with known orbital elements and within 50 pc of the Sun. We see that there

International Workshop: Binaries Inside and Outside the Local Interstellar Bubble
AIP Conf. Proc. 1452, 9-17 (2012); doi: 10.1063/1.4718582

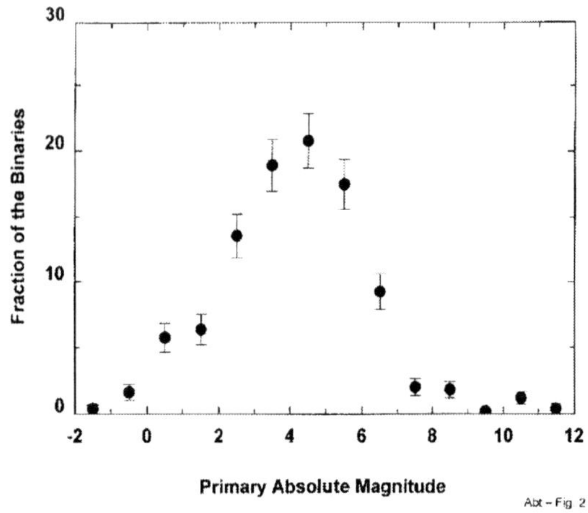

FIGURE 2. The frequencies of 486 VB with known orbital elements in the Sixth Calatog by Hartkopf & Mason and within 50 pc of the Sun.

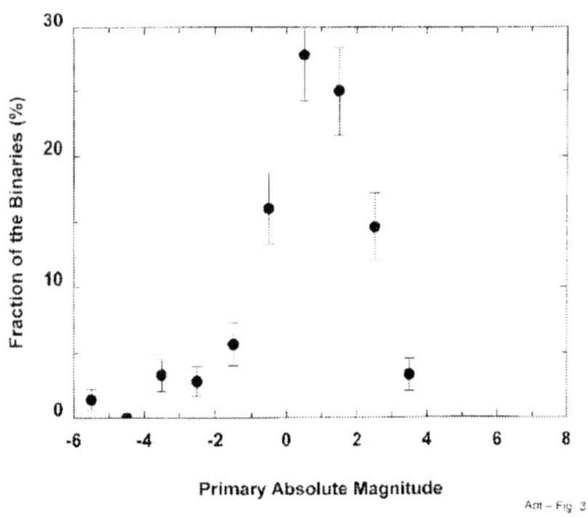

FIGURE 3. The frequencies of 212 VB outside the bubble.

are no stars brighter than B8 or $M_V < -1$. Stars earlier than B7 or B8 leave the main sequence in 160 million years, so the supernova that created that volume must have occurred at about that time. The drop-off after $M_V > 7$ is an observational limitation because only a few such stars (K and M dwarfs) have known orbital elements.

In contrast, Figure 3 shows the absolute magnitudes of 212 VB with known orbital

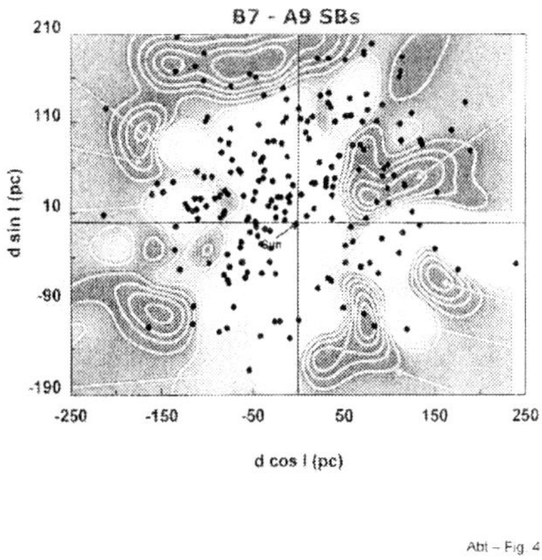

FIGURE 4. The spatial distribution of 212 SB with B7-A9 primaries, taken from the Eighth Catalogue of Batten et al. (1989).

elements and between 150 and 250 pc from the Sun. We see that there are stars that are as bright as $M_V = 6$ or spectral types of O5. So there are many OB stars outside the bubble but none younger than about B8 inside the bubble.

I looked to see if the blast waves from supernovae could disrupt visual binaries with wide separations, but the separations in them do not differ inside and outside the bubble.

Figure 4 shows the positions of 212 spectroscopic binaries (SB) with orbital elements and with primary types of B7-A9. There are many such stars within the bubble. For comparison, Figure 5 shows the positions of 58 SB with orbital elements and with primary stars having types of B0-B6. There are no such stars in the central part of the bubble but there are some in the lobe toward the galactic center and in the Pleiades lobe. This suggest that the central lobe of the bubble is older with only stars of B7 and later, while the lobe toward the galactic center has stars as early as B1 and the Pleiades lobe has stars as early as B3. That would imply that there are three lobes of the bubble that may have been formed by three different supernovae of different ages.

Figure 6 shows the positions of the 100 brightest X-ray binaries within 50 pc of the Sun. They are all within the central part of the bubble, but that is not unexpected because they have spectral types of B8 and later.

Then I looked at the single stars within 150 pc of the Sun. The 21 B7-B9.5 dwarfs (Fig. 7) occur both within and outside the bubble. Looking at the 32 B0-B6 stars (Fig. 8), we see that one B3 star occurs clearly in the Pleiades lobe and four B0-B3 stars fall in the region toward the galactic center. Those confirm the previous results. However there are at least four B2-B6 stars within the central region.

That raises the question whether in tens of millions of years some stars could have drifted into the bubble. For instance, those four stars have an average proper motion

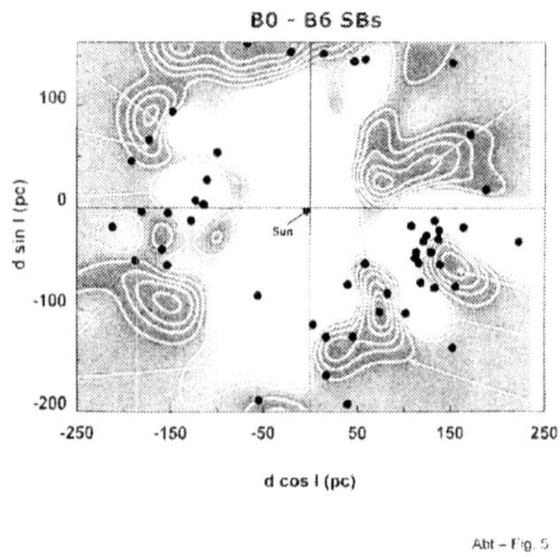

FIGURE 5. The spatial distribution of 58 SBs with B0-B6 primaries.

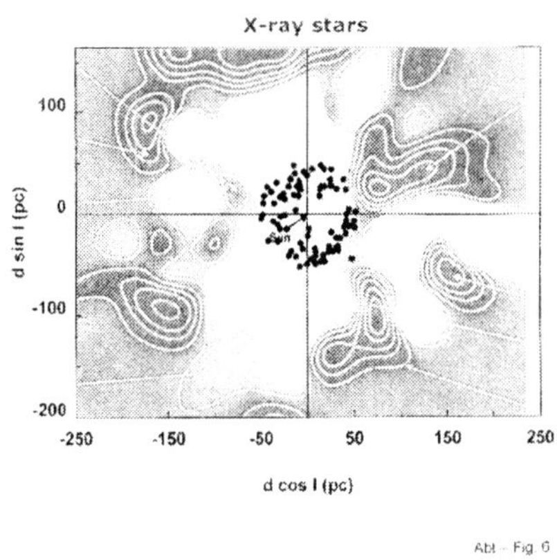

FIGURE 6. The distribution of 100 X-rays stars within 50 pc of the Sun as listed by Makarov (2003).

of 0"050 per year and average distance of 86 Pc. In ten millions years they could have drifted 200 pc, so those stars had plenty of time to drift into the central part of the bubble. Also, the Sun has a motion relative to the galaxy of 11.3 km s^{-1}, so in 100 million years (a minimal age for the supernova that produced the central lobe of the Local Interstellar Bubble), it would have moved 1090 pc from a galactic longitude of 208°. So the Sun

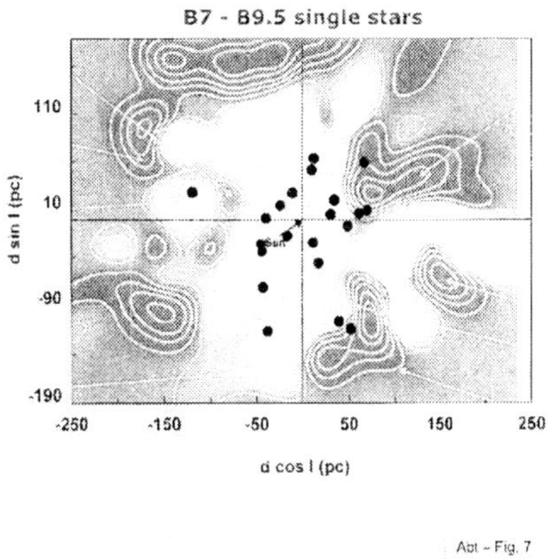

FIGURE 7. The distribution of 21 B7-B9.5 single stars.

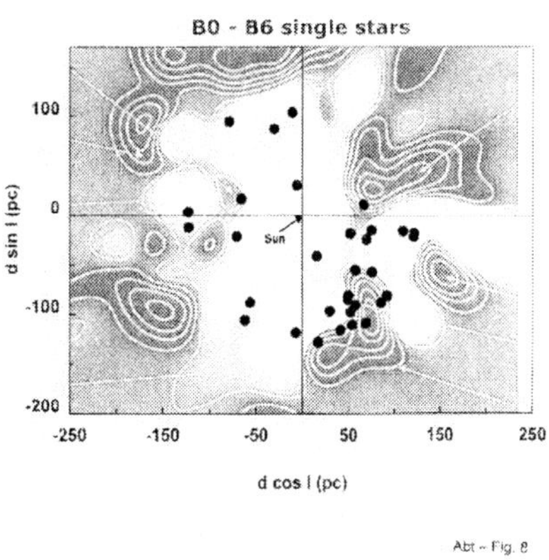

FIGURE 8. The distribution of 32 B0-B6 single stars.

moved into the bubble from outside since the bubble was formed.

We can look at the positions of old nearby open clusters, which are shown in Figure 9. The Pleiades, of course, occurs within the Pleiades lobe. It has stars as early as B6 and an age of about 100 million years. The clusters within the central part of the bubble are old clusters, namely 160 million yr for Collinder 285, 600 million yr for the Hyades,

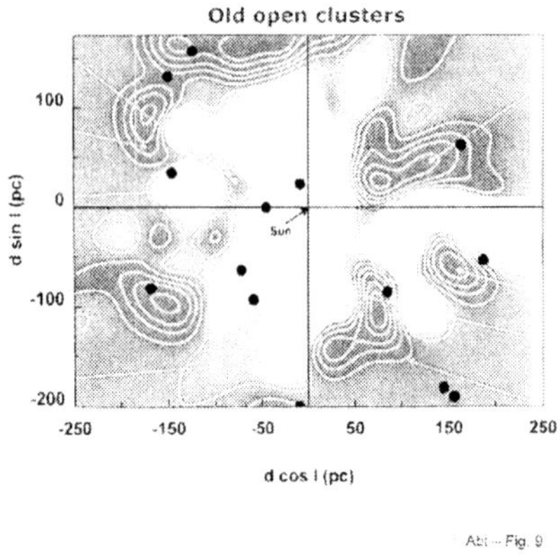

FIGURE 9. The distribution of 14 old clusters (ages > 100 million yr). The cluster at 110° is Collinder 285, at 150° is the Pleiades, at 180° is the Hyades, and at 221° is the Coma cluster.

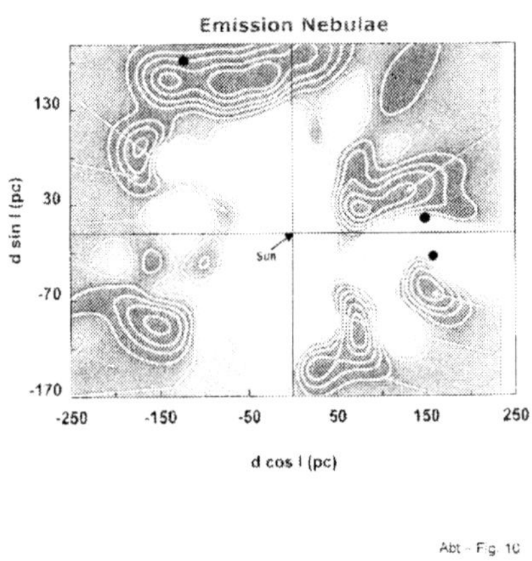

FIGURE 10. The distribution of three emission nebulae.

and 450 million yr for the Coma cluster. There is one young cluster (ρ Oph, not shown) in the lobe toward the galactic center. It has B2 stars and an age of about one million yr.

If the central region of the bubble is old (at least 100 million years), there should be no emission nebulae because those are where stars are formed. Figure 10 shows the three that are within 300 pc. The one (γ Cas) at the top is outside the bubble. The two at the

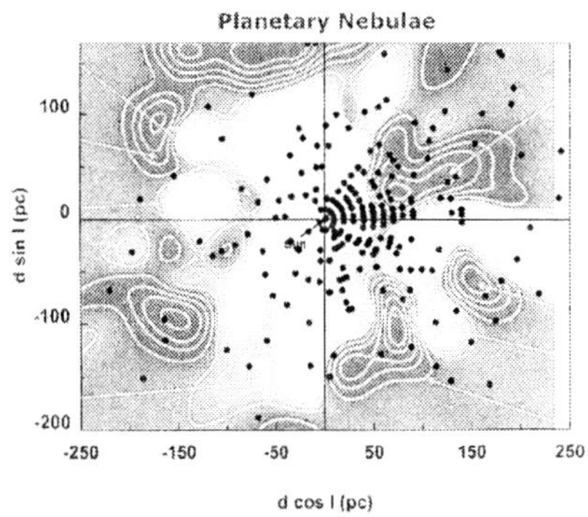

FIGURE 11. The distribution of 279 planetary nebulae as given by Lang (1992). The appearance as lying on circles is due to the fact that their positions are given only to the nearest 10 pc.

right are within the lobe toward the galactic center. Their exciting stars are σ Sco (B2 III) and ζ Oph (O9.5 V); their ages are about 4 million yr. So again we find no emission nebulae or young stars in the central part of the bubble but young stars in the lobe toward the galactic center.

Figure 11 shows the positions of planetary nebulae. Those of 279 planetary nebulae within 250 pc of the Sun lie both inside and outside the bubble. That is not surprising because planetary nebulae originate from 0.8-8 solar mass stars and so they must be at least 500 million years old and predate the supernova explosions that formed the bubble.

Most supernovae leave pulsars as remnants but only a few have radiation jets pointed toward the solar system. Figure 12 shows the galactic positions of the six that lie within 300 pc of the Sun. Four of them are outside the bubble. Pulsar J1856-3754 lies in the lobe toward the galactic center and has a spin-down time of 3.76 million years. That is just the ages of the O9.5 and B0 stars that occur in that lobe. Therefore it seems very likely that the supernova that caused that pulsar is also the one that swept out most of the gas toward the galactic center.

The pulsar J0437-4715 lies just at the bottom edge of the central region, but its spin-down time is 1590 million years, about 10 times as old as the youngest stars in the central region. Therefore it probably predates the formation of the central region. We have no identification for the cause of the Pleiades bubble.

Table 1 summarizes the data from the eight samples of stars. We see that the Pleiades lobe has stars as early as B3, so the age is about 50 million years. The central region of the bubble has stars as early as B7, so its age is about 160 million years. The lobe toward the galactic center has O9.5 and B0 stars, so its age is about 4 million years, which agrees with the pulsar in it that has an age of 3.76 million years.

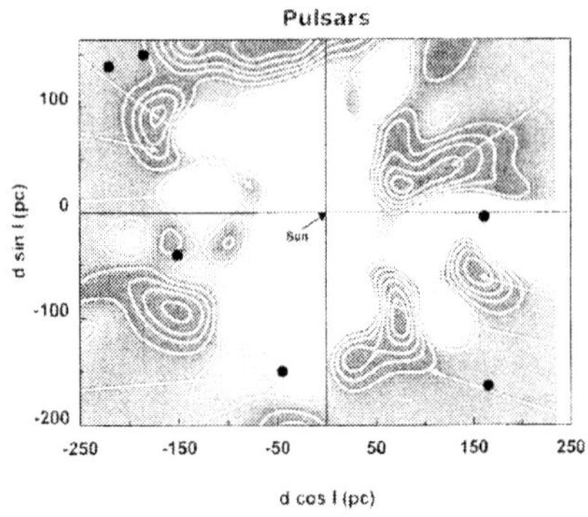

FIGURE 12. The positions of six pulsars within 300 pc (Manchester et al. 2005)

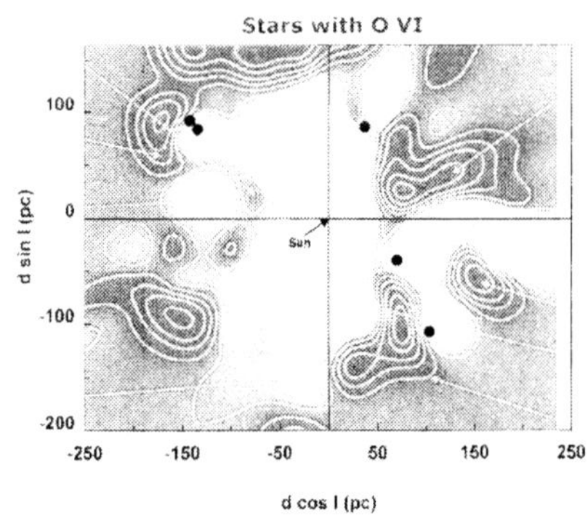

FIGURE 13. The distribution of stars with interstellar 1032 of O VI along their lines of sight (Welsh & Lallement 2008).

If the temperature of the bubble is 10^6 K, we should see virtually no 21 cm H but we should see O VI. Both are true. Welsh & Lallement looked at the 1032 line of O VI and 1036 line of C II with the FUSE satellite. They found those lines to be much stronger than in other parts of the ISM. The locations of five stars with measureable O VI and C II are shown in the Figure 13. The stars are all along the edges of the bubble, so the lines of sight are in the bubble.

It was shown recently (Abt 2004) that B or A dwarfs with disks do not occur within the bubble, although numerous similar stars without disks occur within and outside the bubble. That suggests that the disks are accreted from the interstellar medium and if that medium has a very low density, there are no disks.

TABLE 1. Summary of the Earliest Spectral Types

Sample	Earliest Main Sequence Spectral Types			
	Pleiades lobe	Central region	Toward Gal. Ctr.	Outside
VBs	B8	B8	B8	O stars
SBs	B3	B7	B1	O stars
X-rays	B7	B8	B7	...
Nearby	B3	B7	B0	O stars
Open clusters	B6	B8	B2	O stars
Emision neb.	none	none	O9.5	O stars
Planetary neb.	many	many	many	...
Pulsars	none	none	one	many
Estimated age (10^6 yr)	50	160	3.76	...

REFERENCES

1. Abt, H. A. 2004, ApJL, 603, L109.
2. Batten, A. H., Fletcher, J. M., & MacCarthy, D. G., 1989, Pub. DAO, 17, 1.
3. Lallement, R., Welsh, B. Y., Vergely, J. L., Crifo, F., & Sfier, D. 2003, A&A, 411, 447.
4. Lang, K. R. 1992, in Astrophysical Data: Planets and Stars (New York, Springer).
5. Makarov, V. V., 2003, AJ, 126, 1996.
6. Manchester, R. N., Hobbs, G. B., Teoh, A., & Hobbs, M. 2005, AJ, 129, 1993.
7. Welsh, B. Y., & Lallement, R. 2008, A&A, 490, 707.

Guests in double stars

J. A. Docobo, M. Andrade, P. P. Campo

R. M. Aller Observatory - University of Santiago de Compostela, Box 197, Santiago de Compostela, Spain; joseangel.docobo@usc.es, manuel.andrade@usc.es, pedropablo.campo@usc.es

Abstract. This presentation concerns the study of the movement of Earth-type planets orbiting stars that form binary systems. It is well known that, in accordance with the restricted three-body problem, zones of stability exist either close to the stars or far from both of them, with S-type or P-type orbits, respectively.

We consider the different initial conditions for eight cases (4 of the S-type and 4 of the P-type) and we study the variation of the orbital elements during a moderate time interval, for both the orbit of the binary as well as for the Earth-type planet.

Analyzing the different perturbations, we arrive at the conclusion that in many cases, the existence of Earth-type planets within the zone of habitability is perfectly possible. Only in those cases in which the pair is formed by red dwarfs with P-type orbits, there is a limitation due to the fact that the period of the double star is very small.

We also comment on the possibility that Earth-type exosatellites may exist around giant planets, although it has not been possible to detect the presence of these yet.

Keywords: Celestial Mechanics, Double and Multiple Stars, Extrasolar Planets
PACS: 95.10.Ce, 97.80.-d, 97.82.-j

INTRODUCTION

More than 500 exoplanets (in February, 2011) have been discovered since Mayor and Queloz [8] identified the first one in the mid-90's in the star, 51 Peg. A few years before, an exoplanet was detected in the pulsar, PSR 1257-12, by Wolszczan and Frail [10].

The great majority of the known exoplanets move around single stars. Nevertheless, a significant number of them (more than 40, in February 2011) belong to systems of double and multiple stars.

From the dynamic point of view that concerns the study of movement and stability, this second scenario gives us more possibilities for research and, at the same time, permits us to investigate the possible existence of habitable planets that form part of binary and multiple systems.

In recent years, of course, the study of exoplanets in double stars has been renewed. It should not be forgotten that, a few years before the first of them was discovered, astronomers had already suggested not only their existence but also the possibility that Earth-type planets may orbit the components of a multiple star system [3, 4, 5, 6, 7].

In accordance with the Restricted Three-Body Problem, it is well known that there are specific areas of stability for the movement of infinitesimal bodies. Independent of the Lagrangian points, said areas are close to each of the finite bodies (the stars) or far from both of them.

In this presentation, we want to call attention again to the dynamics of exoplanets but

International Workshop: Binaries Inside and Outside the Local Interstellar Bubble
AIP Conf. Proc. 1452, 18-29 (2012); doi: 10.1063/1.4718583

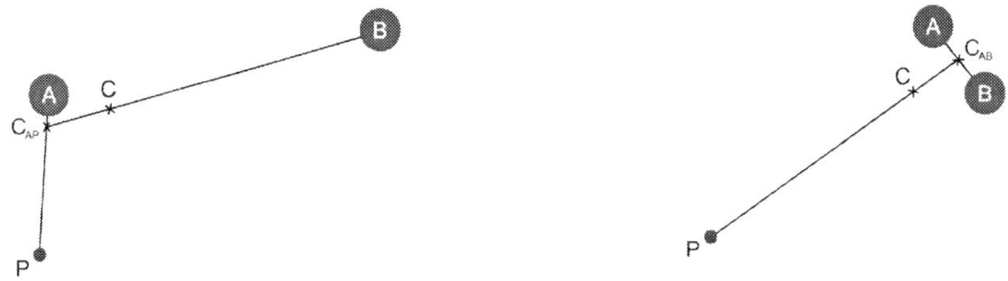

(a) S-type orbit (b) P-type orbit

FIGURE 1. A and B are the components of the double star. P is the planet. C is the center of mass of the system. C_{AP} is the center of mass of A and P (Figure 1(a)). C_{AB} is the center of mass of A and B (Figure 1(b))

also to exosatellites, studying the perturbations of their orbits when they are present in double stars. It is logical to think that there can also be Earth-like satellites around the giant exoplanets and, as such, we consider it to be opportune to also study the dynamics of exosatellites.

EARTH-LIKE PLANETS IN DOUBLE STARS

We will consider the two scenarios proposed by Dvorak [5]: The S-type orbits (Figure 1(a)), and the P-type orbits (Figure 1(b)). In the first case, the planet is moving around a component of a wide binary and, in the second case, the planet is orbiting a close binary.

In both cases, we first suppose that the two stars are similar to the Sun (spectral type G2V and each mass, 1) and then, we will consider the case when the two stars are red dwarfs (e. g., spectral type M0 V and each mass, 0.5). We are going to investigate the long-term evolution of the planet's orbit.

S-type orbits

The following values will be used as the elements of the orbit of the Earth-like planet (hereafter, the angles i, Ω, and ω will be defined with respect to the plane of the binary):

$$
\begin{aligned}
P_E &= 1\ yr. \\
T_E &= 2011.15 \\
e_E &= 0.017 \\
a_E &= 1\ AU \\
i_E &= 5^\circ \\
\Omega_E &= 0^\circ \\
\omega_E &= 45^\circ
\end{aligned}
\tag{1}
$$

If P and a are the orbital period and the semimajor axis of the orbit of the double star, respectively, the following relations should verified:

$$
\begin{aligned}
G(M_1 + M_E) &= \frac{4\pi^2 a_E^3}{P_E^2} \\
G(M_1 + M_2 + M_E) &= \frac{4\pi^2 a_E^3}{P_E^2}
\end{aligned}
\tag{2}
$$

where M_1 and M_2 are the masses of each one of the stars (in this case, $M_1 = M_2 = 1$ solar mass) and M_E, the mass of the Earth-type planet.

We are going to consider two different cases using 30 and 10 AU as the semimajor axes. For each one of these values, the relations (2) give us the corresponding orbital periods: 116.19 years (subcase S1.1) and 22.36 years (subcase S1.2). The remaining orbital elements of the binary will be selected in the following manner:

$$
\begin{aligned}
T &= 2020.0 \\
e &= 0.20 \\
i &= 0^\circ \\
\Omega &= 0^\circ \\
\omega &= 30^\circ
\end{aligned}
\tag{3}
$$

Numerically integrating the equations of the movement of this hierarchical problem of the three bodies (using an implicit Runge Kutta integrator for it) during an interval of 10,000 years, we obtain the results shown in table 1.

Now we suppose, as we commented earlier, that the Earth-type planet is moving around a component of a red dwarf system. The zone of habitability for the planet should be closer to the star with respect to that which it is orbiting because the luminosity of the two components is now less than in the case of solar-type stars. Concretely, the average distance at which the planet should be located in order to receive a quantity of energy similar to that received by the Earth from the Sun, would be close to 0.26 AU This proximity of the planet to the star increases the stability of its orbit which permits the

TABLE 1. Final orbital elements of the planet and the star

Final Orbital elements	Case S1.1		Case S1.2	
	Planet	Star	Planet	Star
$P(yr.)$	0.999887	116.19	1.00413	22.3607
T	2011.1	2020.	2011.39	1997.64
e	0.0308755	0.2	0.0628861	0.2
a (AU)	0.999926	30.	1.00275	10.
$i(^\circ)$	4.99033	0.00000633175	5.08337	0.00000319453
$\Omega(^\circ)$	106.034	36.9335	26.1442	78.9052
ω°	269.521	353.067	205.628	311.099

consideration of situations in which the other component of the binary is also closer than in the previous scenario because the disturbing mass is less as well.

Now, the orbital elements of the planet are going to be the same as those given in (1) except that we take a_E=0.26 In these conditions, if we use

$$\begin{aligned} M_1 &= M_2 = 0.5 \\ M_E &= 1\ Earth\ mass \\ P_E &= 1\ yr. \end{aligned} \tag{4}$$

for $a = 5.59944$ AU, we obtain $P = 13.25$ years, and finally, using $a = 1.8665$ AU, we deduced that $P = 2.550$ years We continue to use the rest of the elements of the binary in the same form as (3). In this manner, we have two new subcases again: S2.1 and S2.2 Integrating the equations throughout the interval of 2,000 years, the results shown on table 2 are obtained.

TABLE 2. Final orbital elements of the planet and the star

Final Orbital elements	Case S2.1		Case S2.2	
	Planet	Star	Planet	Star
$P(yr.)$	0.186983	13.25	0.188508	2.55
T	2011.1	2013.5	2011.12	2019.6
e	0.0187062	0.2	0.0191136	0.2
a (AU)	0.259533	5.59944	0.260942	1.8665
$i(^\circ)$	4.99426	0.00000845193	5.11134	0.00003
$\Omega(^\circ)$	305.68	62.7725	196.21	8.205
ω°	294.51	92.7731	302.57	21.8194

Table 3 illustrates the numeric values of the different perturbations of the e, Ω, and ω elements for the two orbits (planet and binary) and for the four subcases studied in S-type orbits.

Given that the eccentricity only shows periodic perturbations, we indicate the maximum and minimum values along with the time required to pass through two consecutive maximums. Below, in parenthesis, the period of another superimposed perturbation is included. For example, in subcase S1.2, the eccentricity varies between 0.0167 and 0.07 in 980 years with a superimposed motion of 25 years.

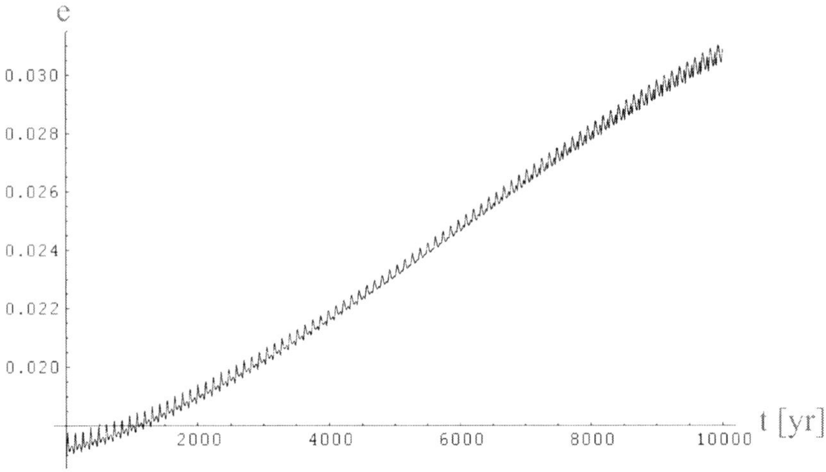

FIGURE 2. Excentricity variation, Case S1.1.

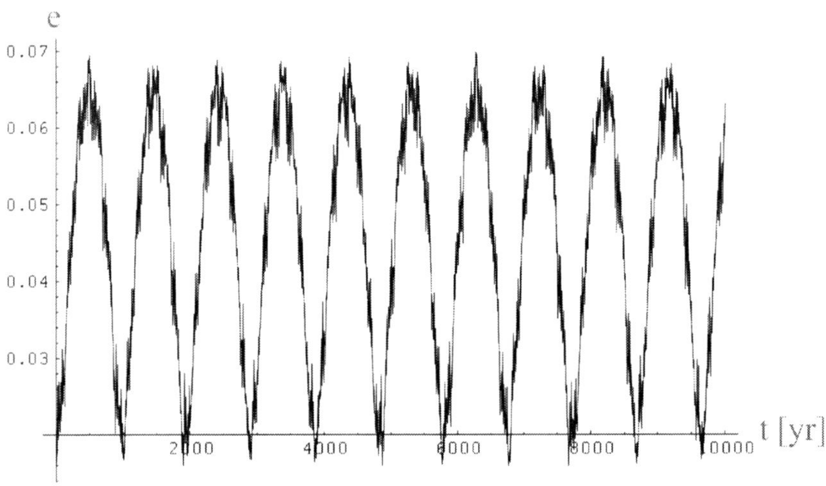

FIGURE 3. Excentricity variation, Case S1.2.

The angle of the node and the argument of the periastron are in circulation in these four cases. We indicate the time necessary for their values to change from 0° to 360°. Immediately below, "secular" is written as this is the principal type of perturbation that occurs. However, a superimposed perturbation of a periodic character also exits; the period appears between parentheses. For example, in subcase 1.2, the node requires 1,250 years in order to make one complete revolution. There is also a periodic motion of approximately 10 years along with the secular perturbation. The arrow indicates if

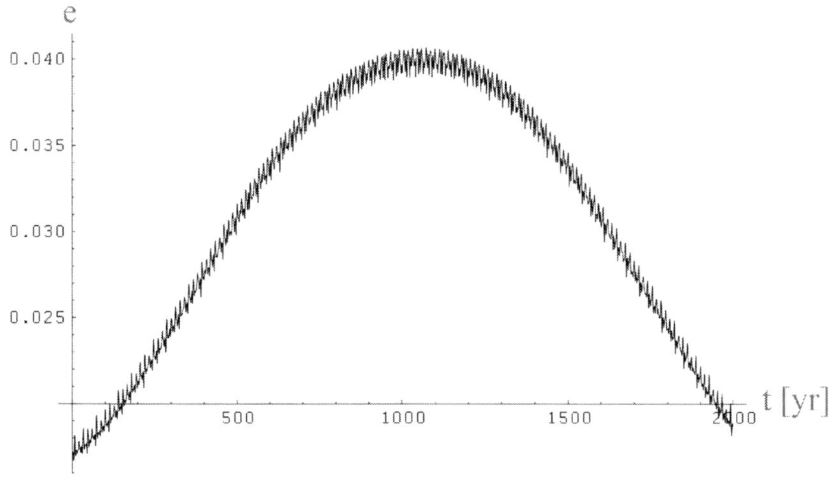

FIGURE 4. Excentricity variation, Case S2.1.

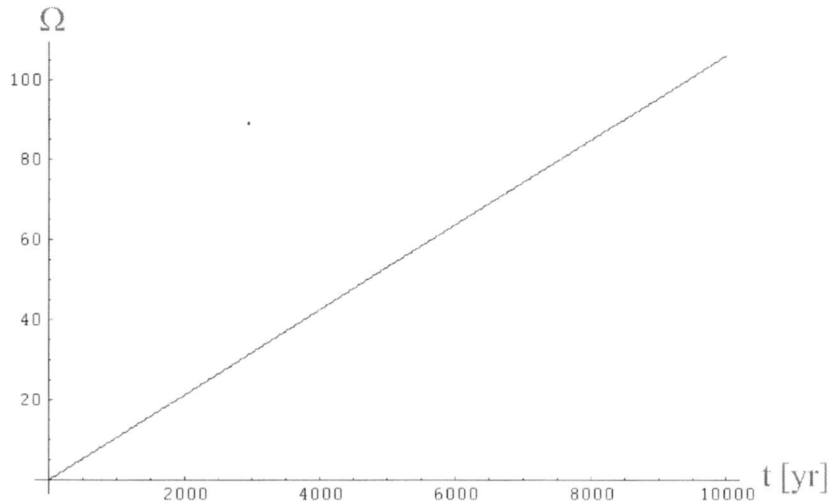

FIGURE 5. Ω variation, Case S1.1.

the variation of the angles is direct or retrograde. The eccentricity of the binary orbit remains practically unchanged (p. u.).

The fourth column in the Table indicates the interval of time used for integration (Δt).

TABLE 3. S-type orbits

SUBCASE	e	Ω	ω	Δt
Planet S1.1	25,000 yr periodic (111 yr.)	34,000 yr. secular ↑ (62 yr.)	17,500 yr. secular ↑ (111 yr.)	10,000 yr.
Binary S1.1	practically unchanged (p. u.)	0° → 36°9 secular ↑	30° → 353°1 secular ↓	10,000 yr.
Planet S1.2	0.0167→0.07 980 yr. (25 yr.)	1,250 yr. secular ↑ (≈10 yr.)	600 yr. secular ↑ (≈15 yr.)	10,000 yr.
Binary S1.2	p.u.	0° → 78°9 secular ↑	30° → 311° secular ↓	10,000 yr.
Planet S2.1	0.0167→0.0400 (13 yr.)	2,360 yr. secular ↓ (12.5 yr.)	1,200 yr. secular ↑ (13 yr.)	2,000 yr.
Binary S2.1	p.u.	0° → 62°8 secular ↓	30° → 92°8 secular ↓	2,000 yr.
Planet S2.2	0.002→0.071 54 yr.	87 yr. secular ↓	31 yr. secular ↑ (0.20 yr.)	2,000 yr.
Binary S2.2	p.u.	0° → 8°2 secular	30° → 21°8 secular	2,000 yr.

P-type orbits

This case corresponds to that illustrated in Figure 2 in which the Earth-type planet moves around the center of mass (C) of the components of a close binary.

Again, considering a first case in which the stars are both similar to the Sun in order to the Earth-type planet to receive a quantity of energy equal to that which the Earth actually receives, it should be situated at a distance close to 1.41 AU from the binary. In that way, we will take the orbital elements as those given in (1), except for the semimajor axis (now 1.41 AU).

Also taking the different values of the semimajor axis of the binary, the expressions:

$$\begin{aligned} G(M_1+M_2) &= \frac{4\pi^2 a^3}{P^2} \\ G(M_1+M_2+M_E) &= \frac{4\pi^2 a_E^3}{P_E^2} \end{aligned} \tag{5}$$

give us the corresponding values of the orbital periods:

a = 0.5 AU P = 0.25 yr. (subcase P1.1)
a = 0.1 AU P = 0.0223607 yr. (subcase P1.2)

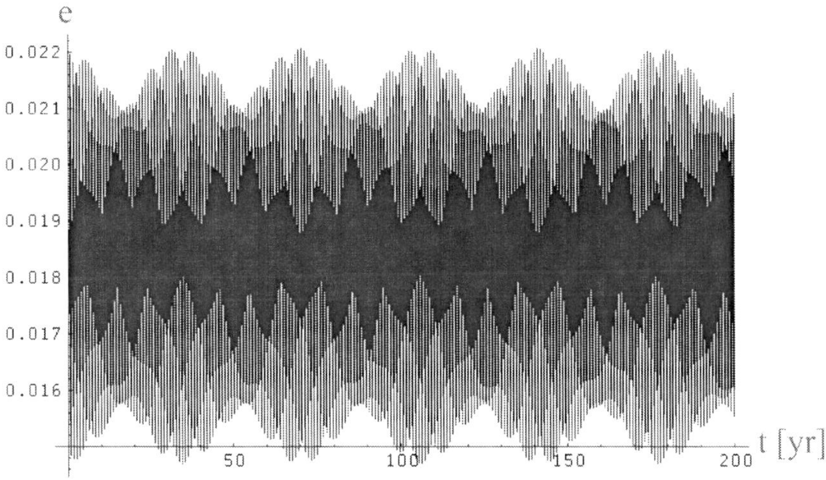

FIGURE 6. Excentricity variation, Case P2.1.

If we consider the close binary formed by two red dwarfs, the semimajor axis of the planetary orbit will now be 0.37 AU and, using again different semimajor axis of the binary, the corresponding orbital periods are now:

a = 0.0464159 AU P =0.01 yr. (subcase P2.1)
a = 0.0092802 AU P =0.000894 yr. (subcase P2.2)

Integrating the equations throughout an interval of 10,000 years in the P1.1 subcase; 1,000 years in the P1.2 subcase; 200 years in the P2.1 subcase, and 20 years in the P2.2 subcase, the following results are obtained:

TABLE 4. Final orbital elements of the planet and the star

Final Orbital elements	Case P1.1		Case P1.2	
	Planet	Star	Planet	Star
$P(yr.)$	1.18354	0.25	1.18859	0.0223607
T	2011.14	2020.0	2019.97	2020.0
e	0.0571258	0.20	0.0174636	0.20
a (AU)	1.40971	0.5	1.41372	0.1
$i(^\circ)$	5.09953	0.000103419	5.03278	0.000106916
$\Omega(^\circ)$	190.051	354.933	304.851	297.227
ω°	290.953	25.3043	289.103	92.7795

Table 6 includes the numeric values of the perturbations. Once again, the eccentricity of the binary orbit remains essentially constant, with only small periodic variations of very low amplitude in relation to the initial value.

TABLE 5. Final orbital elements of the planet and the star

Final Orbital elements	Case P2.1		Case P2.2	
	Planet	Star	Planet	Star
$P(yr.)$	0.224913	0.01	0.225094	0.000894
T	2011.11	2020.	2011.08	2020.0
e	0.0160437	0.199998	0.017087	0.20
a (AU)	0.369836	0.0464159	0.370036	0.00928022
$i(°)$	4.94286	0.000188773	4.97018	0.0000258117
$\Omega(°)$	66.6253	56.1762	3.82318	277.028
$\omega°$	258.115	86.2078	52.7191	307.029

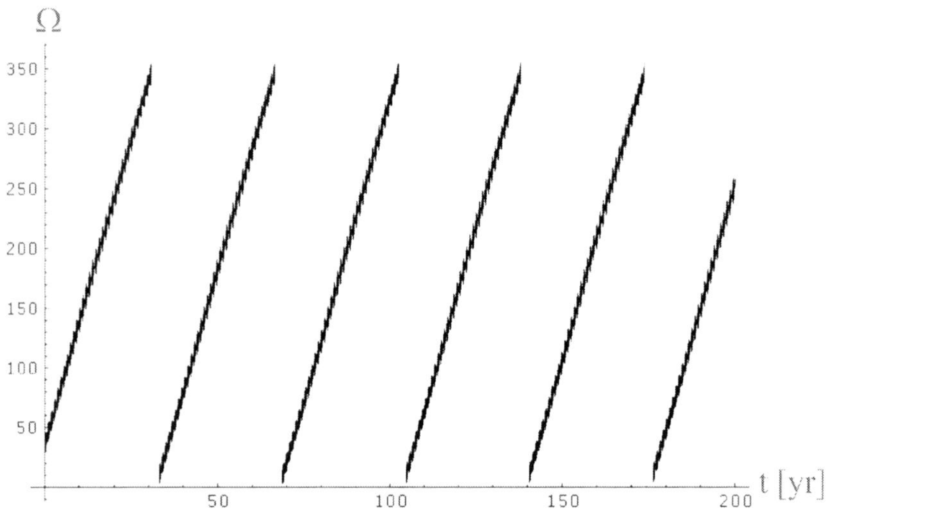

FIGURE 7. ω variation, Case P2.1.

Here it is necessary to point out that in the case of P2.2, the angles Ω and ω, are in libration. That is, they are not going to increase nor decrease secularly as in the previous cases. On the other hand, in order that an Earth-type planet receives the same energy as our planet does from the Sun, it is necessary that the orbital period of the red dwarfs is extraordinarily small. This, in practice, reduces the probability that this type of situation naturally exists.

The eight cases studied in this paper are at the limit of stability in accordance with different criteria. This is so because cases in which the perturbations are significant were specifically selected. Logically, the stability will increase as the disturbing forces decrease, for example, in S-type orbits, considering a great ratio of distances or using lower masses for the distant star or both things simultaneously.

In the case of P-type orbits, it is worthwhile to study the range of masses of the binary for which the existence of stable orbits of Earth-type objects is compatible with

TABLE 6. P-type orbits

SUBCASE	e	Ω	ω	Δt
Planet P1.1	0.0165→0.1100 (333 yr.)	250 yr. secular ↑	267 yr. secular ↑	10,000 yr.
Binary P1.1	practically unchanged (p.u.)	0° → 355° secular ↓	30° → 25° secular ↓	10,000 yr.
Planet P1.2	0.0169→0.0194 600 yr.	600 yr. secular ↓	600 yr. secular ↑	1,000 yr.
Binary P1.2	p.u.	60,000 yr. secular ↓	60,000 yr. secular ↑	1,000 yr.
Planet P2.1	0.0148→0.022 71 yr. (35.5 yr.)	71 yr. secular ↑	36 yr. secular ↑ (≈1 yr.)	200 yr.
Binary P2.1	0.20→0.199998 in 200 yr.	0° → 56° in 200 yr.	30° → 86°2 in 200 yr.	200 yr.
Planet P2.2	0.01691→0.01719 0.23 yr.	4° in 20 yr.	45° → 52°5 in 20 yr. (0.23 yr.)	20 yr.
Binary P2.2	unchanged in 20 yr.	0° → 277° in 20 yr.	30° → 52° in 20 yr.	20 yr.

a reasonable orbital period of the binary. However, in terms of what we have deduced in our work, this will always have to be a spectroscopic binary with a very short period.

In spite of the fact that, until now, the number of planets in P-type orbits was quite small, the Kepler mission has already discovered some others, the majority of which are giant planets.

EARTH-LIKE SATELLITES IN DOUBLE STARS

As occurs in our solar system, the existence of satellites or companions (those with the center of mass of the two bodies situated outside of the planet) moving around the exoplanets may be natural [1, 2]. In our opinion, it is particularly interesting to propose the study of the dynamic evolution and stability of Earth-type exosatellites located in the zone of habitability.

In the case of double stars (and in the most simple situation), all of this leads us to particular cases of the four-body problem. If the detection of exoplanets is difficult, it is much more complicated to detect possible satellites. In fact, none have been discovered yet, not even associate with isolated stars.

The first question that we should ask is if, using conventional methods, we can try to discover the existence of these bodies. For example, employing the direct image is especially difficult today but may be possible in the near future. Photometrically, we would need to be able to achieve very high sensitivity that currently is still not possible. The existence of a satellite orbiting around a planet, both of them moving around a star, should produce variations in the amplitude of the oscillations in the curve of radial

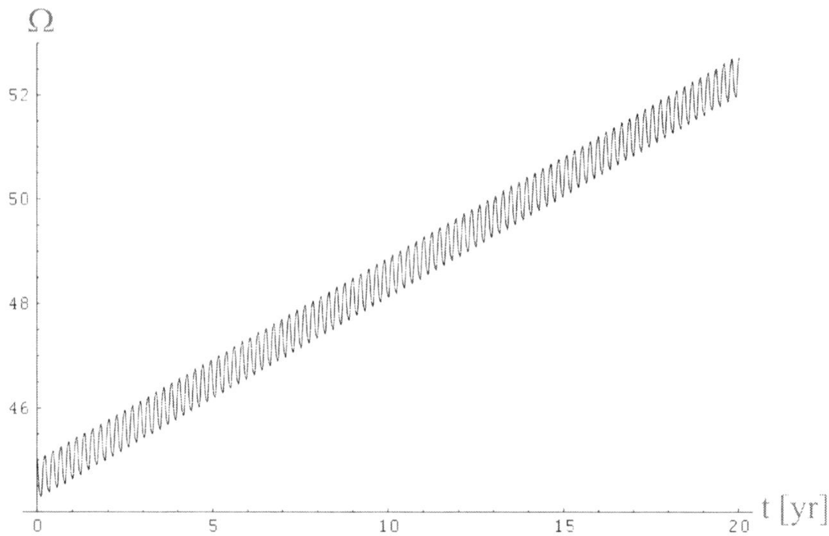

FIGURE 8. ω variation, Case P2.2.

velocity [9]. Nevertheless, if the relation of the satellite/planet masses and/or the mass of the second is small, this theoretical evidence that would lead to the proof of the presence of the satellite is difficult to establish.

The Timing technique could in fact yield results if we study the deviations of the eclipses produced by the passage of the planet in front of the star with high precision. Nevertheless, in the case of binary stars, it is necessary to discriminate between perturbations caused by the satellite and that produced by the distant star.

The perturbation method, that is, the long-term study of the exoplanet orbit variations due to the gravitational action of the satellite, could shed some light on the question. However, in this case, we need very precise orbital elements and adequate software that permit us to distinguish between perturbations caused by exosatellites and other possible exoplanets in the system.

Diverse simulations that we have performed studying the perturbations produced by an exosatellite on the radial velocity curve of the binary with a planet, allow us to prove that the contribution of the exosatellite is on the order of a few cm/sec in the most favourable cases. At the moment, this range is impossible to measure taking into account the current precision of the exoplanet orbits as well as the capacity of the instruments used today.

CONCLUSIONS

As some authors predicted before their discovery, the existence of planets with stable orbits in binary stars is possible, even Earth-type planets.

In fact, the perturbations which may affect them are not an impediment regarding their achievement of stable orbits, according to the cases studied. Neither is it unrealistic to think of the existence of life on exosatellites that orbit giant exoplanets.

Perhaps a great majority of the double and multiple stars that we are studying and whose orbits we are calculating, are nothing more than hosts that receive a large number of guests, including some that capable of supporting life.

REFERENCES

1. Andrade, M.; Docobo, J. A. Satellites around extrasolar planets?. Monografías de la Real Academia de Ciencias Exactas, Físicas, Químicas y Naturales de Zaragoza, n 28, 95-102, 2006.
2. Andrade, M.; Docobo, J. A. *An approach to the dynamics of extrasolar planets with satellites*. Contribution to the "'VII Reunión Científica de la Sociedad Española de Astronomía"', Barcelona, 2006.
3. Dvorak, R. Numerical Investigations of the Development of the Planetary Systems. Mitteilungen. Astron. Gesellschaft Vol. 48, 87, 1980
4. Dvorak, R. Planetary orbits in double star systems. Oesterreichische Akademie der Wissenschaften, mathematisch-naturwissenschaftliche Klasse, Sitzungsberichte, Abteilung 2, vol. 191, no. 10, 423-437, 1982.
5. Dvorak, R. Numerical experiments on planetary orbits in double stars. Celestial Mechanics, vol. 34, 369-378, 1984.
6. Harrington, R. S. Planetary Orbits in Binary Stars. Bulletin of the American Astronomical Society, Vol. 8, 507, 1976.
7. Harrington, R. S. Planetary Orbits in Binary Stars. The Astronomical Journal, vol. 82, 753-756, 1977.
8. Mayor, M.; Queloz, D. *A Jupiter-mass companion to a solar-type star*. Nature, vol. 378, 355-359, 1995.
9. Morais, M. H. M.; Correia, A. C. M. *Stellar wobble caused by a binary system: Can it really be mistaken as an extra-solar planet?*. Astronomy & Astrophysics, vol. 491, 899-906, 2008.
10. Wolszczan, A.; Frail, D. A. *A planetary system around the millisecond pulsar PSR1257 + 12*. Nature, vol. 355, 145-147, 1992.

ANALYZING INFORMATION CONTAINED IN THE OARMA BINARY ORBIT CATALOG

J. A. Docobo, J. F. Ling and P. P. Campo

Ramón Maria Aller Observatory. Universidade de Santiago de Compostela. P.O. Box 197, Santiago de Compostela, Spain

Abstract. This poster displays a statistical summary based on the information contained in the Double Star Orbit Catalog of the Ramon Maria Aller Astronomical Observatory (OARMA). Among other findings, the current aleatory period-eccentricity distribution is particularly interesting. We have correlated eccentricities, periods, spectral types, Hipparcos parallaxes and the orientations of the orbital planes. A list is included that contains all of the names of the authors of the orbits as well as the number of orbits calculated by the authors (February 2011).

Keywords: binaries: visual stars: catalogs
PACS: 97.80 Di

INTRODUCTION

The project of producing the Catalog of Orbits and Ephemerides of Double Stars published by OARMA was first announced by J.A. Docobo in 1996 in his presentation at the International Workshop entitled: "Visual Double Stars: Formation, Dynamics, and Evolutionary Tracks" that was held in Santiago de Compostela, Spain. The idea was to continue the Catalog of Orbits of P. Couteau (inherited by Couteau from P. Muller) but we decided to amplify it by including orbital elements. Almost simultaneously, the Catalog of Orbits prepared by C.E.Worley at the USNO, now managed by B. Mason and W. Hartkopf, also began to include ephemerides and is published on-line. Today, we understand that the two catalogs are complementary and that their co-existence enriches the astronomical information available in this field, although the criteria for inclusion and exclusion are not necessarily the same.

A differentiating concept is related to the manner of evaluating the quality of the orbit. In our case, we do not assign a grade to the orbit; rather, we assign it to the star itself with descriptions of the respective observations and the arc. In this manner, codes A, B, and C correspond respectively to the fact that the binary can have an orbit with a high, average or low quality.

The first edition of the Catalog was published in ACTA ASTRONOMICA (Vol. 51, 353-356, 2001) by J.A. Docobo, J.F. Ling, C. Prieto, J.M. Costa, M.T. Costado, and P. Magdalena. Today it can be consulted on-line at http://www.usc.es/astro/catalog.htm.

Along the years, a sufficient number of orbits (2,122 in February 2011) has been collected which permits us to obtain some statistical results that we believe deserve to be known and diffused.

International Workshop: Binaries Inside and Outside the Local Interstellar Bubble
AIP Conf. Proc. 1452, 30-38 (2012); doi: 10.1063/1.4718584

Concretely, the following information is presented in our poster by figures:

1. Frequency by hemisphere,
2. Frequency by quality code,
3. Frequency of subcomponents and multiple systems,
4. A complete list of the authors along with the number of orbits calculated by them that appear in our Catalog,
5. Frequency by eccentricities,
6. Frequency by orbital periods in years across the spectral types,
7. Frequency by difference of magnitude between the components,
8. Frequency by Hipparcos parallax,
9. The period-eccentricity relationship only for code A and B stars, and
10. Finally, a distribution of the orbital planes with respect to the Equator, graphicaly represented by their angular momentum vector (only for code A stars with the ascending node having been determined).

CONCLUSIONS

From the statistical analysis of the 2,122 orbits studied, we obtained the following results.

1. The distribution of the eccentricities is almost Gaussian with a maximum of frequency close to e = 0.5.
2. Type F, and then G and A, are the more numerous combined spectra but, curiously, no type F appear with orbital periods of less than 5 years.
3. The majority of the visual double stars with calculated orbits have a difference of magnitude between their components of less than 0.5.
4. The Hipparcos parallaxes between 10 and 30 *mas* are the most frequent among the binary stars with orbits.
5. The significant increase in the number of orbits calculated in the last decades shows that the period-eccentricity relation has changed. Currently, no tendency can be deduced when we consider code A and B stars.
6. For the 42 code A orbits with the ascending node determined, results indicate that the correspondent Angular Momentun vector has a negative z-coordinate in 26 cases, and positive in the other 16, when we use an equatorial system of reference.

ACKNOWLEDGMENTS

This work was supported by the Spanish Ministry of Science and Innovation under the Project AYA2010-18433

TABLE 1. Frequency by hemisphere

NUMBER OF STARS: 1,682	
Northern	1103
Southern	579

TABLE 2. Frequency by quality code.

NUMBER OF ORBITS: 2,122	
Code A	266
Code B	421
Code C	976
No code	17

TABLE 3. Frequency of subcomponents and multiple systems.

Spectro/interferometric orbits or Eclipsing/interferometric orbits		60
Orbits with subcomponents	Spectroscopic	53
	Eclipsing	13
Multiple systems		92

TABLE 4. Number of orbits by author/s

Alden	3	Kiyaeva	2
Aller	1	Kiyaeva-Tokovinin-Kalinichenko	2
Alzer-Argyle-Anton	1	Köhler et al.	2
Alzner	20	Kranjc	1
Alzner-Argyle	4	Kranjc-Pigoni	1
Ambruster	1	Lampens et al.	1
Andrade	7	Lampens-Strigachev	2
Arend	1	Lategan	1
Argyle-Alzner	1	Leinert et al.	1
Argyle-Alzner-Horch	3	Ling	24
Aristidi et al.	5	Ling et al.	1
Armstrong et al.	2	Ling-Prieto	6
Artuchyna	1	Lippincott	4
Augensen-Heintz	1	Lippincott-Braun-Mccarthy	1
Baize	93	Lippincott-Turner	1
Balega	5	Liu et al.	1
Balega et al.	10	Luyten.	1
Balega I.-Balega Y.	4	Mante	30
Balega-Bonneau-Foy	1	Martin et al.	1
Balega-Vasyuk	1	Martin-Ianna	1
Barlow-Scarfe-Fekel	1	Mason	10
Barnaby et al.	1	Mason et al.	26
Batten et al.	1	Mason-Douglass-Hartkopf	10
Beardsley	1	Mason-Hartkopf	55
Behall-Harrington	2	Mason-Mcalister-Hartkopf	1
Benedict et al.	1	Mcalister	3
Bennett et al.	1	Mcalister et al.	3
Blanco	1	Morel.	1
Boden-Torres-Latham	1	Mourao	2
Brendley-Hartkopf	14	Muller	7
Brendley-Mason	17	Muterspaugh et al.	3
Brosche	2	Norro	1
Cester	3	Novakovic	25
Chang	1	Novakovic-Todorovic	4
Christy	1	Olevic	20
Churms	2	Olevic-Popovic	1
Cole et al.	1	Olevic-Catovic	1
Costado	1	Olevic-Cvetkovic	28

Costa-Docobo	1	Olevic-Erceg	1
Costa-Morales	3	Olevic-Jovanovic	23
Couteau	40	Osvalds	1
Cvetkovic	30	Pan et al.	1
Cvetkovic-Ninkovic	4	Pan-Shao-Colavita	1
Cvetkovic-Novakovic-Todorovic	10	Pavlovic	1
Cvetkovic-Olevic	1	Pavlovic-Popovic	1
Da Silva-Balca	1	Pavlovic-Todorovic	4
Danjon	1	Peterson-Solensky	4
Docobo	7	Pigulski-Boratyn	1
Docobo et al.	2	Popovic	5
Docobo-Andrade	10	Popovic-Catovic	1
Couteau	40	Popovic-Olevic	1
Cvetkovic	30	Popovic-Pavlovic	11
Cvetkovic-Ninkovic	4	Popovic-Zivkok	1
Cvetkovic-Novakovic-Todorovic	10	Popper-Mcalister	1
Cvetkovic-Olevic	1	Pourbaix	20
Da Silva-Balca	1	Prieur et al.	2
Danjon	1	Rabe	3
Docobo	7	Raghavan et al.	2
Docobo et al.	2	Rakowiecki	1
Docobo-Andrade	10	Rica	17
Dommanget	8	Scardia	37
Dommanget-Arend	1	Scardia et al.	54
Drummond-Christou-Fugote	2	Scardia-Bruzzi	1
Eaton et al.	1	Scardia-Pansecchi	1
Eggen	7	Schaefer et al.	3
Erceg	3	Seifahrt et al.	1
Erceg-Olevic	3	Seymour	3
Fekel et al.	1	Seymour et al.	105
Finsen	13	Seymour-Hartkopf	5
Franz et al.	1	Seymour-Mason	11
Freitas-Debehogne	1	Shakhovskoj et al.	1
Freitas-Mourao	1	Siegrist	1
Gasteyer	1	Simoes Da Silva	1
Gatewood-Han	1	Söderhjelm	77
Geyer-Harrington-Worley	1	Starikova	41
Goldin-Makarov	128	Steffen et al.	1

Gontcharov-Kiyaeva	2	Strand	3
Griffin-Heintz	1	Symms	1
Güntzer-Ligner	2	Tokovinin	10
Hale	12	Tomkin et al.	2
Hall	2	Torres	2
Harrington	1	Torres et al.	1
Harrington-Christy-Strand	2	Torres-Stefanik-Latham	4
Hartkopf	8	Turner et al.	2
Hartkopf et al.	12	Valbousquet	4
Hartkopf-Mason	128	Van Albada	1
Hartkopf-Mason-Mcalister	33	Van Biesbroeck	2
Hartkopf-Mason-Rafferty	15	Van Den Bos	19
Hartkopf-Mcalister-Franz	16	Van Dessel	1
IIcintz	298	Voronov	1
Hershey-Taff	1	Walbaum-Duvent	2
Hirst	1	West	1
Hopmann	17	Wielen	2
Houser	1	Wierzbinski	3
Hummel et al.	11	Wilson	3
Irwin-Yang-Walker	1	Woitas et al.	1
Jasinta.	4	Worley	1
Jastrzebski	1	Zaera	3
Jones	1	Zeller	2
Josties	1	Zirm	47
Kamper-Mcalister-Hartkopf	1	Zirm-Horch	1
Karmel	1	Zivkov-Pavlovic	1
Kiselev et al.	11	Zivkov-Popovic	1
Kiselev-Romanenko-Kalinichenko	18	Zulevic	16
Kiselev-Kiyaeva-Romanenko	2	Zulevic et al.	1

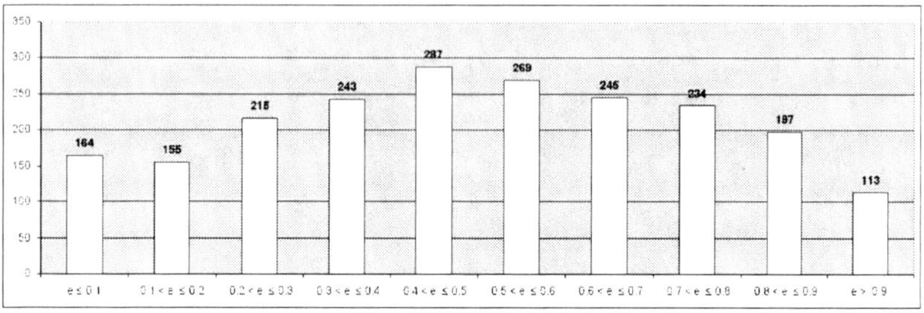

FIGURE 5. Frequency by eccentricities

Sp	P ≤ 5	5 < P ≤ 10	10 < P ≤ 50	50 < P ≤ 100	100 < P ≤ 300	300 < P ≤ 500	P > 500
O	0	0	4	3	6	0	4
B	1	3	15	14	32	15	5
A	5	11	60	40	106	31	55
F	0	20	115	98	227	54	84
G	3	5	93	73	156	47	66
K	7	8	40	44	75	32	40
M	6	0	29	11	7	5	6

FIGURE 6. Frequency by orbital periods in years across the composited spectral types

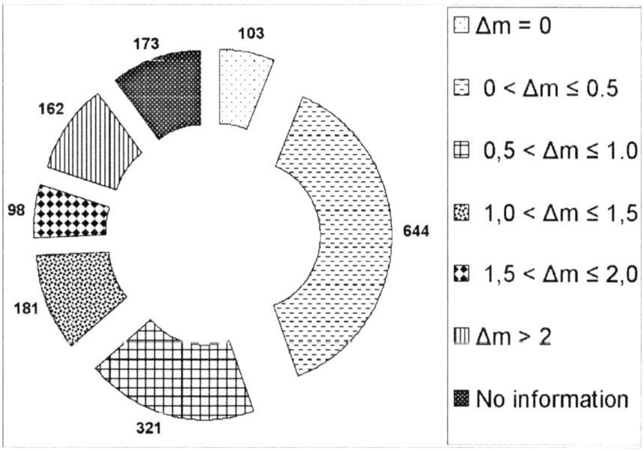

FIGURE 7. Frequency by difference of magnitude between the components

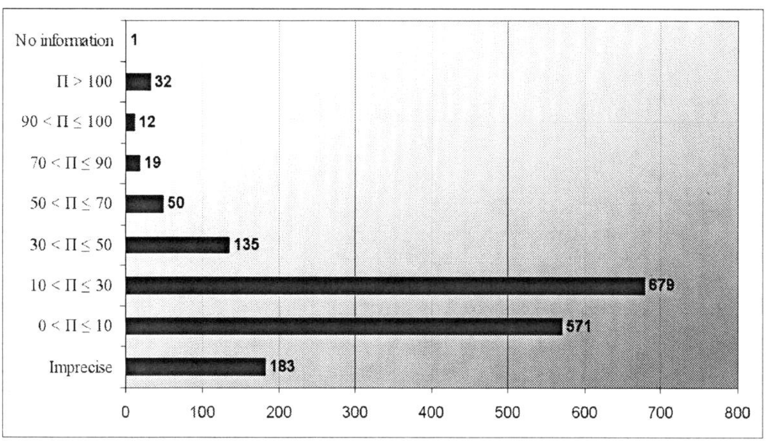

FIGURE 8. Frequency by Hipparcos parallax Π (mas)

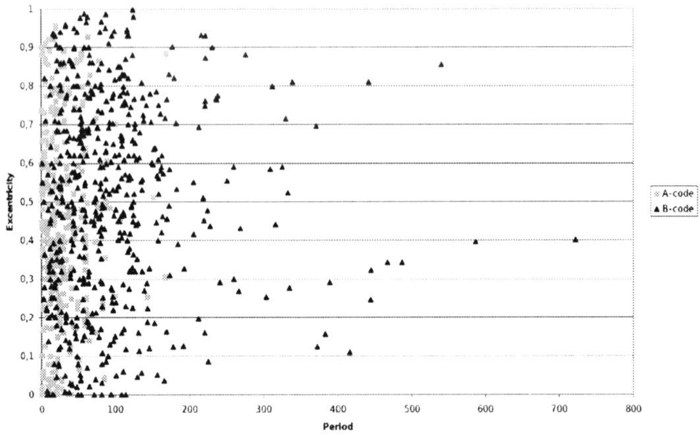

FIGURE 9. The period-eccentricity relationship only for code A and B stars

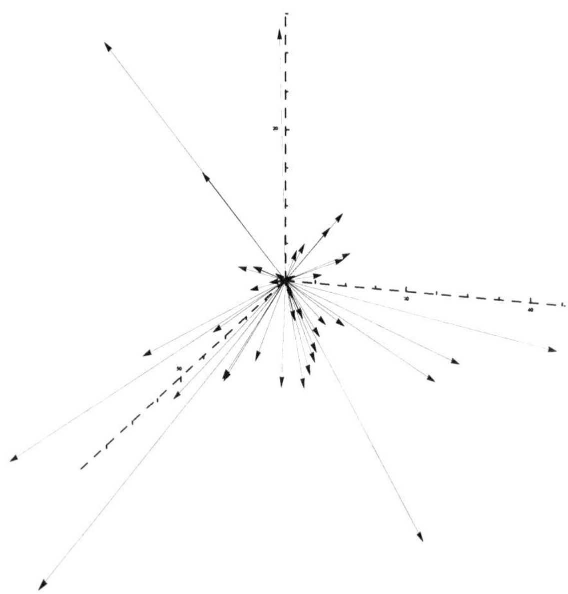

FIGURE 10. Angular momentum vector of the orbits with respect to the Equator

Speckle interferometry with PISCO in Merate and prospects for the future

M. Scardia*, J.-L. Prieur†, R. W. Argyle**, L. Pansecchi*, R. Gili‡ and L. Koechlin†

*I.N.A.F. - Osservatorio Astronomico di Brera, Via E. Bianchi 46, 23807, Merate, Italy
†IRAP CNRS – Université de Toulouse, 14 avenue E. Belin, 31400 Toulouse, France
**Institute of Astronomy, Madingley Road, Cambridge, CB3 0HA, UK
‡Observatoire de la Côte d'Azur, Le Mont-Gros, BP 4229, 06304 Nice Cedex 04, France

Abstract. Speckle interferometry measurements of binary stars have been regularly obtained with PISCO at the 1-meter Zeiss telescope of the I.N.A.F. - Osservatorio Astronomico di Brera (Merate, Italy) since the end of 2003. We present here some recent results and some projects for extending and improving the scientific reach of our program. In particular, we plan to acquire more efficient detectors and install PISCO on a computer-controlled telescope in a better site belonging to Observatoire de la Côte d'Azur (OCA, Caussols, France). In the future, we would like to be able to operate remotely PISCO, its detector and the telescope. Finally we present PISCO2, the new speckle camera of our group that has been built for the OCA 76-cm refractor.)

Keywords: binaries: close, visual – astrometry – techniques: interferometric –
PACS: To be determined

INTRODUCTION

In the previous workshop dedicated to binary stars in Santiago de Compostela (*Double and Multiple Stars: Dynamics, Physics, and Instrumentation*, held in December 2009, we presented a poster about PISCO and its exploitation in Merate ([1]). We invite the reader to consult this publication for a good overview of our scientific program, the methods we use and the most significant results of the exploitation of PISCO in Merate during the period 2004-2009.

In this poster, we successively present a new synthesis of our results the project of our group concerning the possible installation of PISCO on a OCA (Observatoire de la Côte d'Azur, France) 1-meter telescope in Calern and finally our new speckle camera, PISCO2, that is operated on the OCA Nice 76-cm refractor.

International Workshop: Binaries Inside and Outside the Local Interstellar Bubble
AIP Conf. Proc. 1452, 39-49 (2012); doi: 10.1063/1.4718585

PISCO IN MERATE

Brief history

PISCO (Pupil Interferometer Speckle Camera and Coronagraph) is a multi-functional instrument that was initially designed for the 2-meter Telescope Bernard Lyot (TBL) at the Observatoire du Pic du Midi. It was built at the Observatoire de Midi-Pyrénées by J.-L. Prieur and colleagues ([2]) and was first used at the TBL in 1993. Most of the observations were dedicated to speckle interferometry measurements of visual binary stars. The exploitation at Pic du Midi ended in 1999 when the time allocation panel changed its scientific policy and stopped allocating telescope time to PISCO.

A European group of former PISCO users was then created in order to continue the observations of binary stars with PISCO. A first idea consisted of installing this instrument on the Bulgarian Rhozen 2-meter telescope, but various problems led to delaying and finally cancelling this project.

In the beginning of 2003, a new proposition was presented by the Italian members of the PISCO group. Mechanical parts were done during the summer and in October 2003 the instrument was transferred to Merate, some 30 km north-east of Milan where it was fitted to the 1.02-meter Zeiss reflector (Fig. 1). PISCO has been in regular use ever since producing to date more than 2,100 high quality measurements ([6], [7], [9], [11], [3], [13], [4], [14], [5], [15], herein: Papers I to X), from which we derived more than 80 new orbits (see e.g. [12]).

PISCO: a computer controlled multi-purpose instrument

PISCO was designed to be a multi-purpose instrument offering full remote control of all its operating modes (see Fig. 2). With its set of motorized rotating wheels (see Fig. 3), it offers a wide range of possibilities with fast switching between modes (less than one minute) allowing optimal use of the seeing conditions. It is fully computer controlled and all its functions can be selected by a simple mouse click.

Its main features are listed here:

- **Full pupil imaging**: speckle imaging;
- **Masked pupil imaging**: aperture synthesis;
- **Coronagraphy**: imaging with a pupil mask;
- **Spectroscopy**: high angular resolution along the slit;
- **Pupil plane imaging**: SCIDAR (SCintillation Detection And Ranging) or Shack-Hartman with an array of micro-lenses.

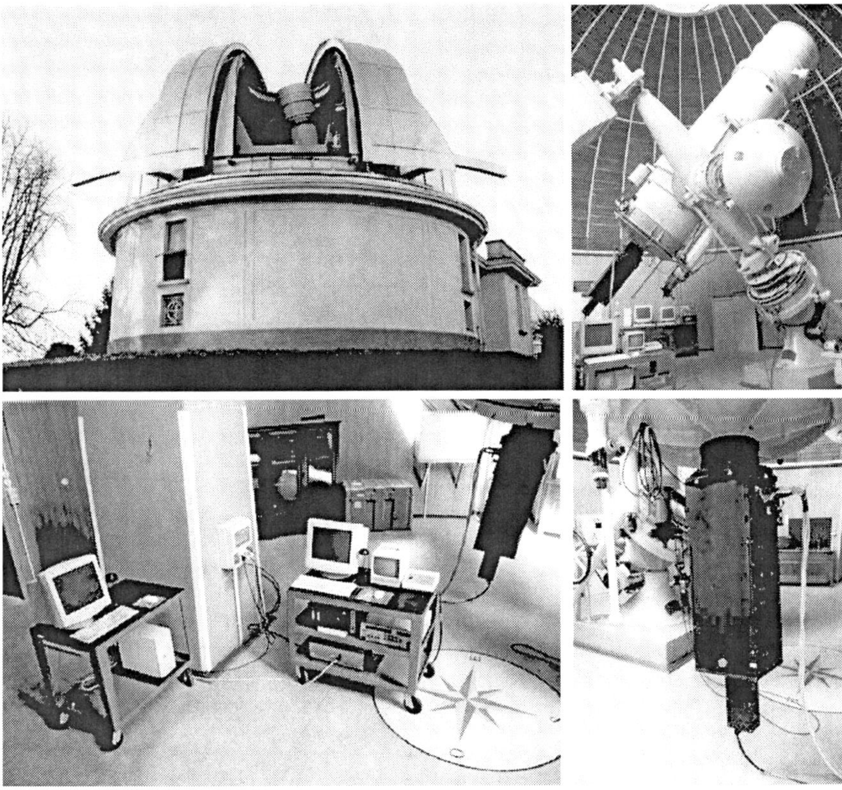

FIGURE 1. PISCO fitted to the 1.02-meter Zeiss reflector in Merate

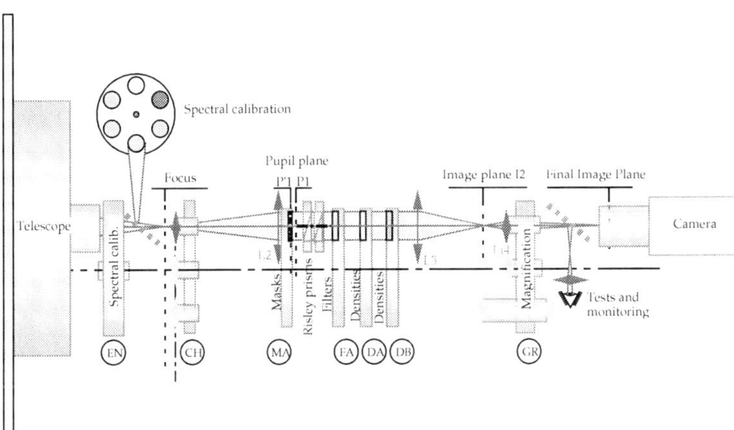

FIGURE 2. The optical layout of PISCO

FIGURE 3. PISCO: detailed view of the inner components.

Scientific exploitation in Merate in 2004-2010

The purpose of our long-term PISCO program is to monitor the relative motion of all the visual binaries accessible with this instrument, fitted to the Zeiss telescope of the I.N.A.F. - Osservatorio Astronomico di Brera in Merate (Italy), for which new measurements are needed to improve orbits. The observing list consists of visual binaries with the following characteristics, which are imposed by instrumental or atmospheric limitations:

(i) declination: north of $-5°$;
(ii) magnitude: brighter than 9.5 - 10.0 in V;
(iii) magnitude difference: less than 4;
(iv) minimum angular separation: $0''.13$ in R;
(v) maximum angular separation: smaller than $\sim 4''.5$.

Limitation (iv) is imposed by the diffraction limit of the Zeiss telescope of 1.02 meter aperture, whilst limitation (v) is chosen so that the binary systems fit inside the isoplanatic patch of the atmosphere, which is a necessary condition for speckle measurements.

Our scientific program consists mainly in obtaining good-quality measurements of those visual binaries in order to determine accurate orbits and derive stellar masses. For this purpose PISCO is tuned in its full pupil image mode and speckle interferometric techniques are used to derive measurements of binary stars. The observing procedure and the data reduction methods are described in detail in [1]

Since 1st January 2004, more than 2,100 high quality measurements of visual binaries have been made at Merate with typical accuracies of around $0''.01$ and $0°.6$, with a limiting separation close to the diffraction limit of the telescope. The results have been discussed in ten papers (Papers I to X) and allowed us to calculate 81 new or improved orbits.

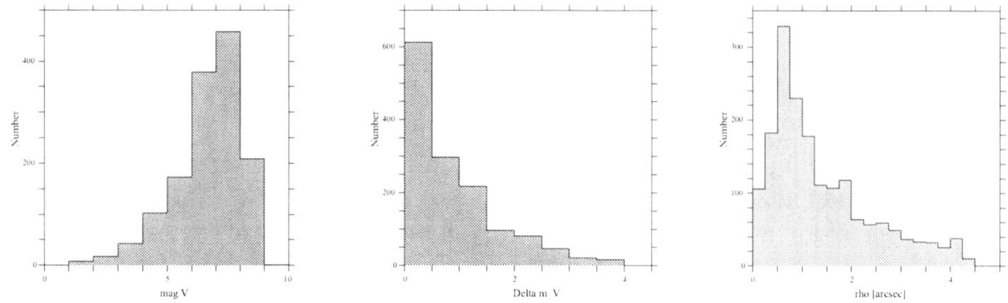

FIGURE 4. Objects measured in 2004–2009: visual magnitude (left), difference of magnitudes of the two components (center) and angular separations (right). The total number of measurements exceeded 2100 and concerned more than 1700 objects (see Papers I to X).

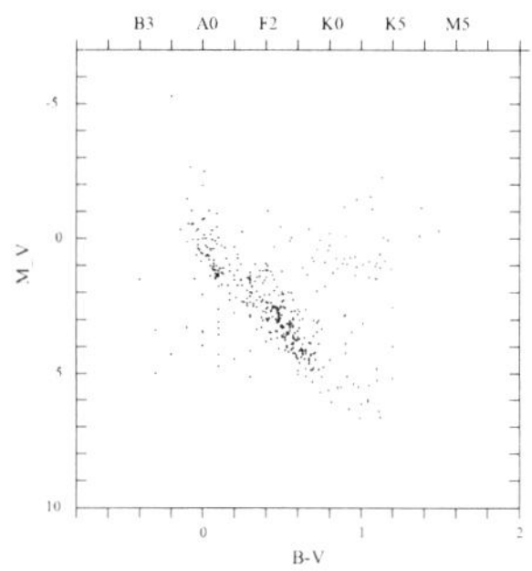

FIGURE 5. Objects measured in 2004-2009: HR diagram built with the objects for which the Hipparcos parallax error was less than 20%.

The distribution of the angular separations is displayed in Fig. 4 and shows a maximum for $\rho \approx 0''.7$, with a full range between $0''.1$ and $4''.6$. Some measurements were even obtained below the diffraction limit $\rho_d = \lambda/D \approx 0''.13$ with the R filter (i.e. $\lambda = 650$ nm) and the Zeiss telescope whose aperture is $D = 1.02$ m. The smallest separation was measured for WRH 28 with $\rho = 0''.066 \pm 0.012$ (Paper IV [11]).

The distribution of the apparent magnitudes m_V of the binaries measured between 2004 and 2009 and the difference of magnitudes Δm_V between the two components are presented in Fig. 4. The telescope aperture and detector sensitivity lead to a limiting magnitude of about $m_V = 9.5$. A careful handling of the image processing enabled us to measure a few couples with $\Delta m_V > 3$. The distribution of the errors on ρ (angular separation) shows that typical accuracy is around $0''.01$. For the position angle, it is about $0^\circ.6$.

Using the Hipparcos parallaxes, we were able to construct the HR diagram of those binaries that is displayed in Fig. 5. It can be seen that a large part of the HR diagram is covered by our sample. In the future we would like to acquire a more sensitive detector in order to observe fainter (and cooler) main sequence stars.

PISCO IN CALERN

The situation around the Merate observatory has been gradually degrading in the last few years. Indeed it is located in the outskirts of the small town of Merate which has increased the number and intensity of its street lights. Furthermore big accommodation complexes are being built just across the street, in front of the observatory, and in the Southern side of the Zeiss telescope dome.

We are therefore investigating the possibility of moving PISCO to another better site. We have received a good reception of this project from OCA Observatory. On the Calern plateau (see Fig. 6), two 1-meter telescopes are being restored (see Fig. 7) and one of them could be used for our project.

Moving PISCO from the Merate Zeiss telescope to one of those two 1-m telescope in Calern has many advantages:

- Modern telescope: it will be more reliable and will allow automatic pointing of the objects;
- Better site: the night sky background is less luminous and the average seeing conditions are better.
- Higher altitude and better transparency: possibility of observing objects with lower elevation.

Furthermore, we hope to be able to acquire a more sensitive detector than our old ICCD detector, with an expected quantum efficiency of about 60%, which is ten times larger than that of our ICCD (note though that the LUCA and IXON EMCCD detectors that we have tested in 2009 and 2010 with PISCO now were very disappointing and did no better than our ICCD). It will also offer the advantage of being computer-controlled, and the output data will be directly digitized by the camera (which is not the case with our analog ICCD detector).

We should be able to fully control the observations remotely. Indeed PISCO is fully computer-controlled and the elementary frames are processed in real-time with our software. With a new detector and a modern telescope that will also be computer-controlled, we will be able to control all the phases of the binary observations from a computer. Although we think that a technician will always be needed to check that everything runs smoothly, but the observation of binary stars with PISCO may be done without the presence of an astronomer and should significantly gain in productivity.

The combination of a better site and a more sensitive detector should allow us to observe faint red stars such as objects in the lower end of the main sequence, and pre-main sequence stars. Those stars are too faint to be observed with our current instrumentation in Merate, as shown in the HR diagram plotted in Fig. 5. Observing them and deriving their orbits would become a priority of PISCO observations since the determination of the masses of young stars is important to constrain star formation models.

FIGURE 6. Calern observatory site in summer (left) and winter (right).

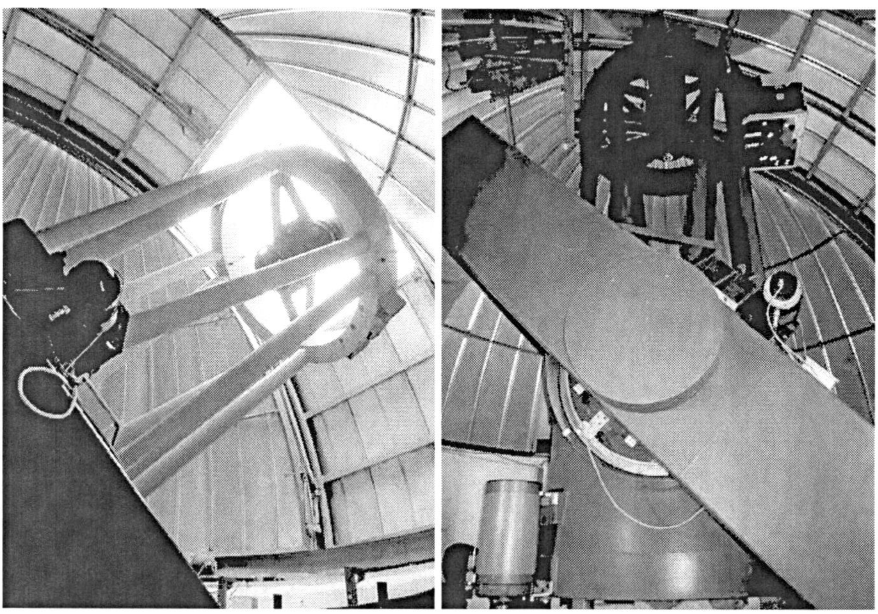

FIGURE 7. One of those two 1-m telescopes at Calern that may be used for binary stars observations with PISCO.

TABLE 1. Theoretical resolution limit of the telescopes and filters used by our group for observing visual binaries.

Filter	B	V	R	J	H	K	
λ_C (μm)	0.450	0.550	0.650	1.253	1.643	2.177	
λ_C/D for D=0.7 m	0.13	**0.16**	0.19	0.37	0.49	0.64	Nice
λ_C/D for D=1.0 m	0.093	**0.11**	**0.13**	0.26	0.34	0.45	Merate
λ_C/D for D=2.0 m	0.046	**0.057**	**0.067**	0.13	0.17	0.22	TBL
λ_C/D for D=3.6 m	0.026	0.032	0.037	**0.072**	**0.094**	**0.12**	ESO

PISCO2 IN NICE

A new speckle camera, PISCO2, was built in 2009 for the Nice 76-cm refractor to observe visual binaries using speckle interferometry techniques (see Fig. 8). Due to a tight budget, its optical layout is a simplified version of that of PISCO (see Fig. 2), with fewer filters and no pupil mask. The implementation of Risley prisms for the correction of the atmospheric dispersion will be done in the future (hopefully in 2012).

The diffraction limit of this refractor is slightly larger than that of 1-m telescopes (see Table 1), but the quality of the images is very good with typical seeing values smaller than one arc second. As a result very faint binary stars can be observed with PISCO2 in Nice, fainter than with PISCO in Merate.

The detector used with PISCO2 is a back-illuminated ANDOR IXON EMCCD with a good sensitivity. In particularly good conditions, we have observed binaries of $m_V = 15$.

We have developed a new program `buildspeck1` that controls ANDOR-type cameras and offer the same real-time image processing possibilities as the program `vcrb` that is used in Merate with our ICCD camera. An example of recent observations of ADS 3174 (= STF 535) is shown in Fig. 9). The EMCCD gain was set to 4095, its temperature was regulated at $-70°$C and the processed sequence had 1150 frames. One can see the quality of the short exposure (lower right frame), where the two stars are clearly separated.

FIGURE 8. PISCO2 on the 76-cm Nice refractor

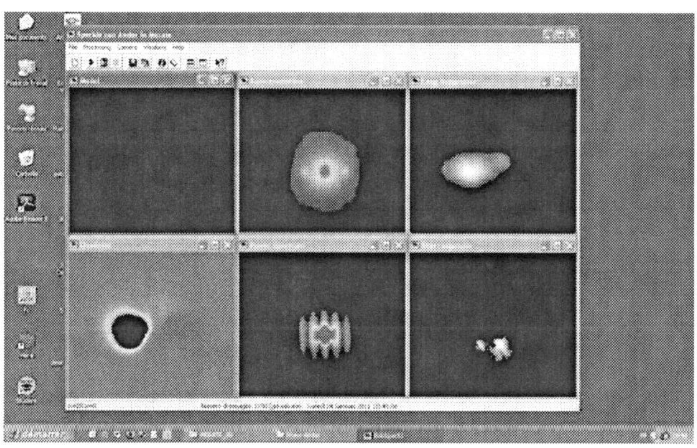

FIGURE 9. ADS 3174 (=STF 535) observed on Jan. 24th 2011 with PISCO2 and the EMCCD ANDOR SPICA camera in Nice (rho=1".1, theta=270°). From left to right and top to bottom: mean autocorrelation, long integration, restricted triple-correlation, power spectrum, and elementary frame.

CONCLUSION

Our program with PISCO in Merate has shown that PISCO is a reliable, transportable, easy-to-use instrument, which has led to more than 2100 measurements with good astrometric accuracy (0".01, 0°.6). The project has also already led to the determination of the orbits and mass sums of many visual binaries. More than 80 orbits were computed by our team using PISCO observations. The accuracy of the masses found by will be largely improved when the distances are accurately known (Hipparcos was insufficient).

The Zeiss telescope has been for many years the largest telescope in the world dedicated to the observation of visual binary stars. The availability of a dedicated telescope is essential for studying binary stars since it allows orbital monitoring during periastron passage.

The PISCO team is now looking forward to improve both the efficiency of the detector and to moving our instrument to a better site, so that fainter pairs can be observed i.e. those towards the bottom of the Main Sequence. Negotiations are underway with Observatoire de Côte d'Azur to use PISCO on a 1-meter telescope at the Plateau de Calern in south-eastern France. We also want to replace our ICCD detector with a new EMCCD detector. The combination of these two factors should increase the magnitude limit available to PISCO by about 3 magnitudes allowing the measurement of closer and fainter binaries in particular those at the lower end of the main sequence.

With a modern telescope and a new detector, we should be able to make automatic observations of visual binaries with PISCO, remotely controlled by a computer. We think that our initiative is likely to be generalized in the world and that remote-controlled observations will become the rule for measuring visual binary stars in the future.

With PISCO at Calern and PISCO2 at Nice, we will be in a good position to significantly increase the number of binary stars measurements from which new orbits and stellar masses will be derived. This program would benefit from the precise parallaxes that the GAIA space mission should provide in the near future, and we will thus contribute to the worldwide project of creating of a large data base of stellar masses.

REFERENCES

1. Argyle R.W., Scardia M., Prieur J.-L., Pansecchi L., 2011, *Double and Multiple Stars: Dynamics, Physics, and Instrumentation*, held 10-11 December 2009, in Santiago de Compostela (Spain) Eds: J.A. Docobo, V.S. Tamazian, Y.Y. Balega ISBN 978-0-7354-0902-6 2011, AIP Conf. Proc. 1346, 42–56
2. J.-L. Prieur, L. Koechlin, C. André, C. Gallou & C. Lucuix, *E*xperimental Astronomy, **8**, 297 (1998)
3. J.-L. Prieur, M. Scardia, L. Pansecchi, R.W. Argyle, M. Sala, M. Ghigo, L. Koechlin, E. Aristidi, *M*NRAS, **3**87, 772 (2008) (Paper V)
4. J.-L. Prieur, M. Scardia, L. Pansecchi, R.W. Argyle, M. Sala, *M*NRAS, **3**95, 907 (2009) (Paper VII)
5. J.-L. Prieur, M. Scardia, L. Pansecchi, R.W. Argyle, M. Sala, *M*NRAS, **4**07, 1913 (2010) (Paper IX)
6. M. Scardia, J.-L. Prieur, M. Sala, M. Ghigo, L. Koechlin, E. Aristidi, F. Mazzoleni, *M*NRAS, **357**, 1255, (2005) (Paper I)
7. M. Scardia, J.-L. Prieur, L. Pansecchi, R.W. Argyle, M. Sala, M. Ghigo, L. Koechlin, E. Aristidi, *M*NRAS, **3**67, 1170 (2006) (Paper II)

8. M. Scardia, R.W. Argyle, J.-L. Prieur, L. Pansecchi, S. Basso, N.M. Law and C.D. Mackay, *A*N, **3**28, 2, 146 (2007)
9. M. Scardia, J.-L. Prieur, L. Pansecchi, R.W. Argyle, S. Basso, M. Sala, M. Ghigo, L. Koechlin, E. Aristidi, *M*NRAS, **3**74, 965 (2007) (Paper III)
10. M. Scardia, J.-L. Prieur, L. Pansecchi, R.W. Argyle, S. Basso, N.M. Law and C.D. Mackay, Proceedings IAU Symposium 240, 132+558 (2007)
11. M. Scardia, J.-L. Prieur, L. Pansecchi, R.W. Argyle, M. Sala, S. Basso, M. Ghigo, L. Koechlin, E. Aristidi, *A*N, **3**29, 1, 54 (2008) (Paper IV)
12. M. Scardia, J.-L. Prieur, L. Pansecchi, R.W. Argyle, *A*N, **3**29, 4, 379 (2008)
13. M. Scardia, J.-L. Prieur, L. Pansecchi, R.W. Argyle, M. Sala, *A*N, **3**30, 1, 55 (2009) (Paper VI)
14. M. Scardia, J.-L. Prieur, L. Pansecchi, R.W. Argyle, M. Sala, *A*N, **3**31, 286 (2010) (Paper VIII)
15. M. Scardia, J.-L. Prieur, L. Pansecchi, R.W. Argyle, M. Sala, *A*N, **3**32, 508 (2011) (Paper X)

"Visual" Orbit Solutions from Observing Techniques Old and New

Brian D. Mason
U.S. Naval Observatory

ABSTRACT

For the past couple of decades, the primary double star observational technique utilized at the US Naval Observatory has been speckle interferometry. Our two speckle cameras have been used on a variety of telescopes: our 26″ in Washington, the USNO (NOFS) 61″, the McDonald 82″, the Mt. Wilson 100″ and the twin 4 meter telescopes of KPNO and CTIO. While these instruments have each yielded many observations, they have been involved in rather different programs, with the 26″ observing primarily the so-called "neglected" pairs and larger instruments conducting duplicity surveys and observing close, astrophysically interesting systems.

While speckle is quite successful at resolving relatively close pairs, for those pairs which are both bright and very close long baseline optical interferometry may be the only viable solution. Utilizing both the CHARA Array and the Navy Prototype Optical Interferometer, numerous hard-to-observe pairs first resolved by speckle have been observed using these arrays' superior resolution capabilities; this has allowed these orbits to be significantly improved; examples of pairs observed by each of these instruments are presented.

At the other separation extreme, a cache of photographic plates taken with the USNO double star camera has been digitized and processed. Some 66 plates of Sirius A and B taken between 1970 and 1984, which represents a 10% increase in the total number of measures, have been reduced, enabling a significant improvement over the current "best" Sirius orbit, almost one full revolution later.

1. 26″ Speckle Observing

Due to its modest aperture, our capability to observe orbital motion pairs of astrophysical signicance is quite limited. As a result, in 2000 the program shifted over to pairs that were unconfirmed or had not been observed in a decade or more. These we defined as "neglected". The eventual goal is the characterizing of pairs as either optical or physical.

This definition was adopted by many others and has formed a productive operational observing program for us and other groups allowing the re-invigoration of observing programs with limited access to large telescopes. To date we have obtained 24,707 mean positions with our speckle camera on the 26″ telescope in Washington. As a measure of our success observing these neglected pairs (as well as others with similar programs), the number of mean positions per system in the WDS has increased from 5.8 to 7.1 despite the number of new systems ($n = 1$) increasing by 36,852 pairs.

Historically, when we obtained access to larger telescopes and observed off-base, 26″ observing would shut down for approximately a month. Due to the success of of our 26″ neglected doubles program, a second camera for use with our older ICCD was constructed in-house (Mason et al. 2007) and our primary camera has since been used when we have had access to larger telescopes, simultaneous with continued 26″ operation with our secondary camera.

International Workshop: Binaries Inside and Outside the Local Interstellar Bubble
AIP Conf. Proc. 1452, 50-55 (2012); doi: 10.1063/1.4718586

2. Large Telescope Speckle Observing

For various programs we have observed with our speckle camera on many different telescopes. These telescopes include the Naval Observatory Flagstaff Station (NOFS) 61″ telescope (neglected pairs, see Hartkopf & Mason 2008), the McDonald Observatory 82″ telescope (Hipparcos doubles, see Mason et al. 2001), the Mt. Wilson Observatory 100″ telescope (neglected pairs, see Hartkopf & Mason 2009), and the 4m telescopes of Kitt Peak National Observatory and Cerro Tololo Interamerican Observatory (massive stars, see Mason et al. 2009). Other data are in process.

These programs include astrophysically interesting orbit pairs, surveys for new companions, fainter and southern hemisphere pairs. These programs have resulted in 3608 observations, 89 new pairs, and 260 orbits. Additional observations have also been obtained with the SOAR 4.2m telescope with HRCam (Tokovinin et al. 2010).

An example of an individual system of interest over a long period of time was given the memorable moniker "Tweedledum and Tweedledee". This pair of nearly identical interferometric pairs was discovered by Finsen (FIN 332, 1953). The similarity in brightness, separation and orientation led to many issues of misidentification as well as orbital quadrant ambiguities. It has been under regular observation by speckle inteferometry since the 1970s, most notably by CHARA and USNO. Older unpublished CHARA observations and USNO speckle observations from all the above large telescopes were included in an analysis of both pairs resolving the long-standing quadrant issues (Mason et al. 2010a). Investigation also indicates they are within 1σ of being co-planar.

Additional orbits of southern interferometric pairs have also been presented (Figure 1; Mason et al. 2010b) and others are in preparation. For these figures, the dashed curve is the previous best orbit. The filled stars are USNO observations published since the time of the dashed curve, filled circles are other observations since then, and the solid curve is the new orbit based on all data.

3. Navy Prototype Optical Interferometer

The Navy Prototype Optical Interferometer (NPOI) is located on Anderson Mesa, outside Flagstaff, Arizona. See Zavala et al. (2010) for an example of NPOI double star work. The current operational baseline allows routine observing to a resolution of 1 mas. Ultimately this capability will be $\sim$200 μas. This R to V band instrument has a current limit of V $\sim$6. This will be extended significantly when the former Keck Interferometer outrigger 1.8m telescopes are integrated.

Mason et al. (1997) published combined solution orbits of three hot dwarf & cool giant pairs: HR 233, 36 Tau and 73 Leo. While the single lined RV spectroscopy and deconvolved composite spectra are quite good, the resolved measures were hampered by the inability to resolve the pair at all orbital phases, even with speckle interferometry on the largest accessible telescopes. While long baseline optical interferometry solves this problem, the magnitude difference (Δm) in the infrared, where interferometers work best, generates other problems. The V band operation of the NPOI makes it ideal for these targets. See Figure 2.

Other hot dwarf & cool giant pairs are also under investigation. One of the longest period eclipsing systems, γ Per, is very well characterized in one quadrant, but has no observations in the other. From 2019.7 to 2021.1 this pair is expected to be at less than 30 mas separation and will spend part of that time in the opposite quadrant. Coverage here will characterize the orbit quite well. Another rarely resolved challenging system is τ Per (McAlister 1981). This cool giant & hot dwarf pair is ideal for the capabilities

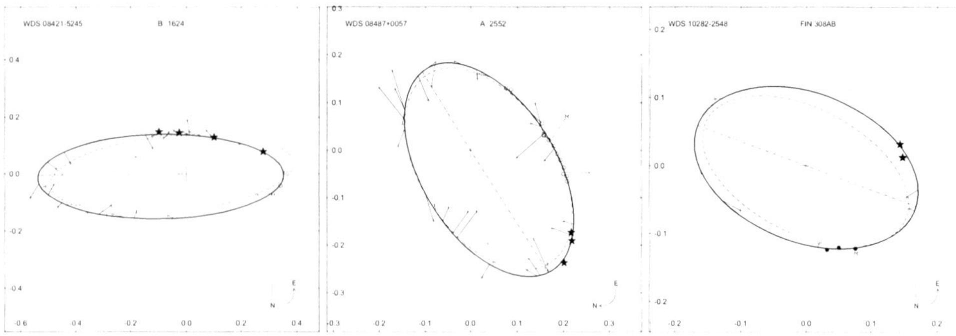

Fig. 1.— New orbital solutions, plotted together with all published data in the WDS database. In each of these figures, micrometric observations are indicated by plus signs, modern interferometric measures by filled circles, and older eyepiece interferometry measures by open circles; Hipparcos measures are indicated by the letter 'H'. The new measures enabling these orbital improvements are indicated as filled stars. "$O-C$" lines connect each measure to its predicted position along the new orbit (shown as a thick solid line). Dashed "$O-C$" lines indicate measures given zero weight in the final solution. A dot-dash line indicates the line of nodes, and a curved arrow in the lower right corner of each figure indicates the direction of orbital motion. Finally, the previous published orbit is shown as a dashed ellipse. These orbits are from (left to right) Mason et al. (1999), Hartkopf & Mason (2000) and Docobo (1991).

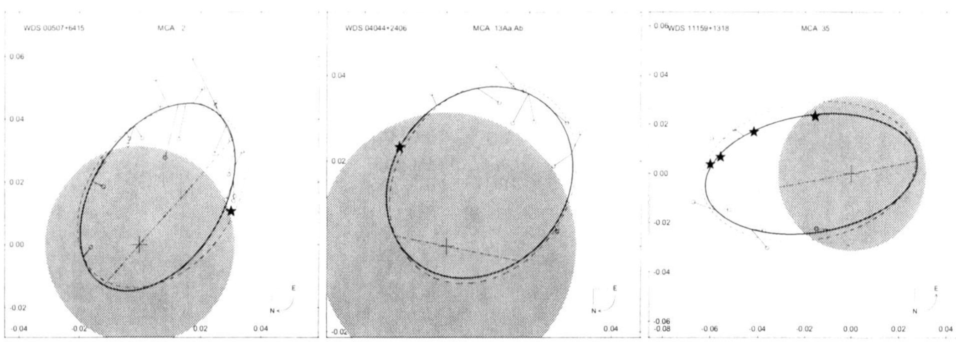

Fig. 2.— New orbital solutions, plotted together with all published data in the WDS database as well as unpublished NPOI observations. In each of these figures, modern interferometric measures are plotted as open circles and the filled stars indicate the NPOI measures. The large shaded circle centered on the origin is the 30 mas V band resolution limit of speckle interferometry on a 4m telescope. The previous orbits (dashed curves) are from Mason et al. (1997). Other symbols are as in Figure 1. Observation of these pairs are ongoing. New orbits are courtesy of USNO summer intern Haley Hurowitz.

of the NPOI. In the case of δ Sge the pair is a hot dwarf and a cool supergiant. The orbit of Eaton et al. (1995) had no reliable observations due west at the closest separations. Successful NPOI observations of all these pairs have been obtained, and observations continue.

4. CHARA Array

The CHARA Array, on Mt. Wilson in California, is an interferometer similar in configuration to the NPOI. The emphasis on achieving results on the longest baselines in infrared makes this instrument capable of higher resolution than the NPOI at present. However, most results are pairwise and give the interferometer observables of baseline and visibility. Combining multiple baseline pairs allows triangulation of the binary through analysis of their separated fringe packets (SFP). This methodology is thoroughly explored in Farrington et al. (2010).

While HD 178911 is observable for much of its orbit with single aperture interferometry, the improved accuracy and precision of SFP observations, plus the capability to resolve at less than 30 mas, have made this pair very attractive. Publication of the SFP solution is expected following the 2011 observing season. The pair μ Ari has a SFP measure in the hitherto unresolved region due south and to the west. The 2011 observable wedge here is due north in a region not well covered before. See Figure 3.

A recent application of multiple systems is to use the wider single component in a multiple system as the calibrator of the close binary. In this case, the wider pair is a known speckle binary appropriate for speckle interferometry and SFP solutions. An orbital analysis of both systems from speckle interferometry, SFP work, baseline/visibilty interferometry data as well as spectroscopy has recently been presented (ten Brummelaar et al. 2011). This orbit is also plotted in Figure 3.

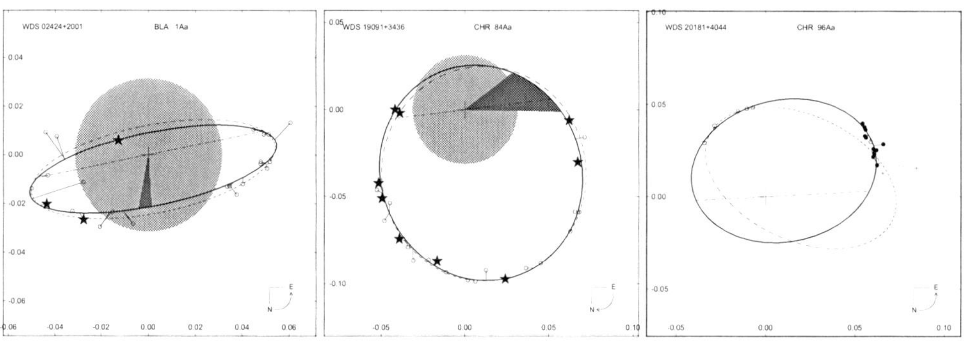

Fig. 3.— New orbital solutions, plotted together with all published data in the WDS database as well as unpublished CHARA observations. In each of these figures, published interferometric measures are plotted as open circles and filled symbols indicate the CHARA SFP measures. The large shaded circle centered on the origin is the 30 mas V band resolution limit of speckle interferometry on a 4m telescope. The previous orbits, indicated as a dashed curve, are from (left to right) Mason (1997), Hartkopf et al. (2000) and Hartkopf et al. (1993). For the left and center figures, the shaded wedge indicates where the companion is expected to be in Fall 2011 when the pairs are scheduled for observation. Here filled stars indicate CHARA SFP measures obtained over the last several years which are unpublished. For the pair at right, the published SFP measures are plotted as filled circles. The orbit is the new speckle+SFP+RV determination of ten Brummelaar et al. (2011). Other symbols are as Figure 1.

5. Photographic Data: Sirius

At the other separation extreme, a cache of photographic plates taken with the USNO double star camera has been digitized and processed, using the method described by Lindenblad (1970). Some 66 plates of Sirius A and B taken between 1970 and 1984, which represents a 10% increase in the total number of measures, have been reduced. They enable a significant improvement over the current "best" Sirius orbit (van den Bos 1960), almost one full revolution later. The orbit is presented in Figure 4.

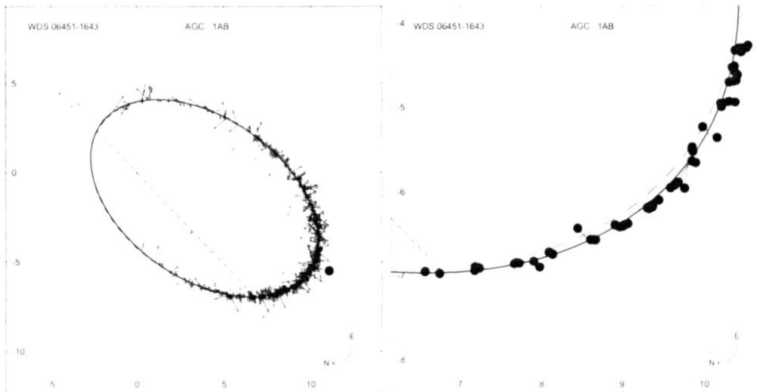

Fig. 4.— New orbital solutions for Sirius. In the figure at left all published data in the WDS database as well as unpublished USNO photographic observations (filled circles) are plotted. In the figure at right a 4× zoom of the lower right plots only the newly reduced observations, the new orbit as a solid curve and the current "best" orbit (van den Bos 1960) as a dashed curve. New orbit are courtesy of USNO summer intern Miranda Seitz-McLeese. Final calibration of these data is being performed with the assistance of USNO retiree Jerry Josties.

Currently, the technique utilized most frequently for systems with large magnitude difference is adaptive optics. We are also exploring continued work in this area (Turner et al. 2008).

6. Conclusions

While we continue to utilize speckle interferometry as our primary observational technique, the parameters of separation, magnitude and differential magnitude require a plethora of methods each of which covers different portions of $\rho - m - \Delta m$ space. We will continue to seek out partnerships with other groups to address them.

Thanks are extended to Ken Johnston and the USNO for their continued support of the Double Star Program.

REFERENCES

ten Brummelaar, T.A., O'Brien, D.P., Mason, B.D., Farrington, C.D., Fullerton, A.W., Gies, D.R., Grundstrom, E.D., Hartkopf, W.I., Matson, R.A., McAlister, H.A., McSwain, M.V., Roberts, L.C.,

Jr., Schaefer, G.H., Simon-Diaz, S., Sturmann, J., Sturmann, L., Turner, N.H. & Williams, S.J. 2011, AJ 142, 21

van den Bos, W.H. 1960, JO 43, 145

Docobo, J.A. 1991, Inf. Circ. 115

Eaton, J.A., Hartkopf, W.I., McAlister, H.A. & Mason, B.D. 1995, AJ 109, 1856

Farrington, C.D., ten Brummelaar, T.A., Mason, B.D., Hartkopf, W.I., McAlister, H.A., Raghavan, D., Turner, N.H., Sturmann, L., Sturmann, J. & Ridgway, S.T. 2010, AJ 139, 2308

Finsen, W. S. 1953, Mon. Not. Astron. Soc. Southern Africa 12, 86

Hartkopf, W.I. & Mason, B.D. 2000, Inf. Circ. 142

Hartkopf, W.I. & Mason, B.D. 2008, AJ 135, 1334

Hartkopf, W.I. & Mason, B.D. 2009, AJ 138, 813

Hartkopf, W. I., Gies, D. R., Mason, B. D., Bagnuolo, W. G., Jr., & McAlister, H. A. 1993, BAAS, 25, 872

Hartkopf, W.I., Mason, B.D., McAlister, H.A., Roberts, L.C., Jr., Turner, N.H., ten Brummelaar, T.A., Prieto, C.M., Ling, J.F., & Franz, O.G. 2000, AJ 119, 3084

Lindenblad, I. 1970, AJ 75, 841

Mason, B.D. 1997, AJ 114, 808

Mason, B.D., Douglass, G.G. & Hartkopf, W.I. 1999, AJ 117, 1023

Mason, B.D., Hartkopf, W.I., Gies, D.R., Henry, T.J. & Helsel, J.W. 2009, AJ 137, 3358

Mason, B.D., Hartkopf, W.I., Holdenried, E.R. & Rafferty, T.J. 2001, AJ 121, 3524

Mason, B.D., Hartkopf, W.I. & McAlister, H.A. 2010a, AJ 140, 242

Mason, B.D., Hartkopf, W.I. & Tokovinin, A. 2010b, AJ 140, 735

Mason, B.D., Hartkopf, W.I., Wycoff, G.L. & Wieder, G. 2007, AJ 134, 1671

Mason, B.D., McAlister, H.A., Hartkopf, W.I., Griffin, R.F. & Griffin, R.E.M. 1997, AJ 114, 1607

McAlister, H.A. 1981, AJ 86, 1397

Tokovinin, A., Mason, B.D. & Hartkopf, W.I. 2010, AJ 139, 743

Turner, N.H., ten Brummelaar, T.A., Roberts, L.C., Jr., Mason, B.D., Hartkopf, W.I. & Gies, D.R. 2008, AJ 136, 554

Zavala, R.T., Hummel, C.A., Boboltz, D.A., Ojha, R., Shaffer, D.B., Tycner, C., Richards, M.T. & Hutter, D.J. 2010, ApJ 715, 44

Eclipsing binaries: observational data, classification and parameterization

O. Malkov

Institute of Astronomy of the Russ. Acad. Sci., 48 Pyatnitskaya St., Moscow 119017, Russia
Faculty of Physics, Moscow State University, Moscow 119992, Russia

Abstract. Eclipsing binaries represent the most numerous type of binaries with observable orbital period. Independent stellar mass and luminosity determination is possible only for components of eclipsing binaries, with the spectrum lines of the two components. However, they make up only about 2% of all known eclipsing binaries. So, for statistical investigations it would be advisable to estimate fundamental parameters (such as mass and radius) for eclipsing binaries with unknown spectroscopic elements.

In this paper we briefly discuss modern catalogues and lists of eclipsing variable stars. Then we describe methods of determination of astrophysical parameters of binary components from a limited set of observational data and present results of application of the methods to existing lists of eclipsing variables.

Keywords: Binary stars, eclipsing binary stars, catalogues, databases
PACS: 97.80.Hn, 95.80.+p

1. INTRODUCTION

The mass-luminosity relation, MLR, is a fundamental law that is used in various fields of astrophysics. For single objects, the MLR allows astronomers to convert a relatively easily observed quantity, luminosity, to a more fundamental characteristic, mass, that yields a better understanding of the object's nature. It is especially important for the construction of the initial mass function from the luminosity function of stars.

Independent stellar mass and luminosity determination for intermediate mass ($1.5 < m/m_{\odot} < 30$) stars is possible only for components of eclipsing binaries with the spectrum lines of the two components. Eclipsing binaries (hereafter EB) are very numerous and can be discovered at large distances (in particular, EB is the only type of binaries that can be observed in other galaxies). However, only some 2% of all known EB display the spectrum lines of both components.

Also, a lot of questions concerning evolution and structure of contact binaries are still open. Some types of EB can serve as an indispensable tool for solving these problems.

So, it is advisable to elaborate methods to estimate at least approximate values of fundamental parameters (such as mass and radius) for EB with unknown spectroscopic elements. Obviously, the methods may differ for EB of different evolutionary types.

That was a goal of authors of early studies (Brancewicz & Dworak [1980], Svechnikov & Kuznetsova [1990]). However, the number of known eclipsing binaries increases constantly: current version of the Catalogue of Eclipsing Variables (CEV) contains about 7200 eclipsing binaries, in comparison with some 6300 systems in the 2006 version (Malkov et al. [2006]). The number of eclipsing variables discovered in large

International Workshop: Binaries Inside and Outside the Local Interstellar Bubble
AIP Conf. Proc. 1452, 56-72 (2012); doi: 10.1063/1.4718587

surveys like OGLE, MACHO, ASAS-3, etc. is much larger. So it is advisable to estimate parameters of eclipsing binaries with the most modern data.

In this paper we present a procedure that allows us, after classification of an eclipsing binary, to estimate fundamental parameters of its components with a minimum input data: period, spectrum of primary, relative radii and evolutionary class. It significantly enlarges the number of systems, which the procedure can be applied to. Also, new approaches were developed for parameterization of some types of eclipsing binaries.

Catalogues and lists of eclipsing variable stars are briefly discussed in Section 2. Basic parameterization procedure and its application to different types of eclipsing binaries is discussed in Section 3. Finally, in Section 4 we draw our conclusions.

2. MODERN CATALOGUES AND LISTS OF ECLIPSING BINARY STARS

Construction of catalogues / databases of binary stars has a long history. Compilations of data for eclipsing binary stars have been published by Brancewicz & Dworak [1980], Wood et al. [1980], Svechnikov & Kuznetsova [1990], Shaw [1994], Maceroni & van't Veer [1996], Gettel et al. [2006], and others.

A new catalogue of 6330 eclipsing variable stars, chosen from the General Catalogue of Variable Stars, GCVS (Samus et al. [2011]), was compiled by Malkov et al. [2006]. This Catalogue of Eclipsing Variables (CEV) represents the largest list of eclipsing binaries classified from observations. Beside that, thousands of eclipsing binaries are included in microlensing (OGLE, MACHO, EROS, MOA) and all-sky photometric (ASAS) surveys. A list of more recent catalogues of various types of eclipsing binaries is compiled and published in Kaygorodov et al. [2012].

3. DETERMINATION OF FUNDAMENTAL STELLAR PARAMETERS OF ECLIPSING BINARY SYSTEMS

3.1. Basic equations

The following basic expressions can be written for an eclipsing system and its components. Components of binaries should satisfy the Stefan-Boltzmann law

$$L = R^2 T^4, \tag{1}$$

and a system should satisfy the third Kepler law

$$\frac{a^3}{P^2} = \frac{G}{4\pi}(m_1 + m_2). \tag{2}$$

Here L, R, T, m are luminosity, radius (called also absolute radius), effective temperature and mass of a component, correspondingly, hereafter expressed in solar unit; a is a semi-major axis of the system in solar radius, and P is its period in year. Absolute radius of a component can be calculated from

$$R = ar, \tag{3}$$

where relative radii r of components are derived from the binary light curve.

If luminosity class (LC) of a component is known, than L, T, m can also be estimated from its two-dimensional spectral type Sp:

$$L = L(Sp), \quad T = T(Sp), \quad m = m(Sp). \tag{4}$$

In the present study we use detailed de Jager & Nieuwenhuijzen [1987] spectrum-temperature and spectrum-luminosity tables for T and L estimation.

As to the latter equation in (4), mean stellar masses according to spectral type and luminosity class were presented by Dworak [1975]. To compile a more detailed spectrum-mass relation (Table 1 and Fig. 1) we have used empirical data from stellar mass catalogues: Belikov [1995], Malkov [1993], Kraicheva et al. [1980] (only dynamical masses were used from the latter catalogue). Our LC=V spectrum-mass relation is essentially Svechnikov & Taidakova [1984] main sequence relation, slightly extrapolated to earlier spectra with our empirical data. Spectrum-mass relation from Kraicheva et al. [1980] was used for bright giants later than A0II, due to lack of other empirical data in this region. Also, there is no reliable empirical data for supergiants B5I to G4I in the literature.

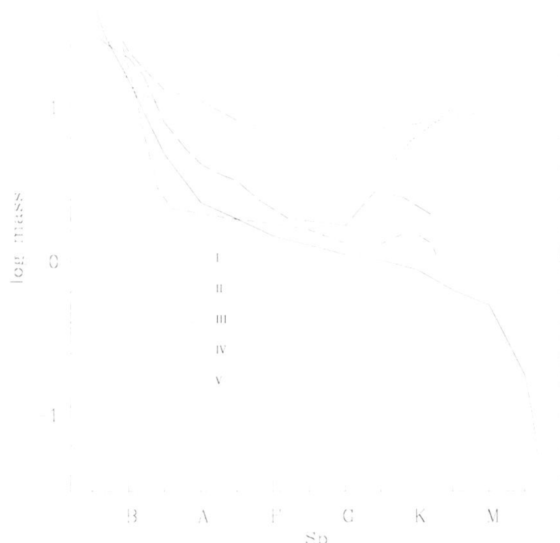

FIGURE 1. Spectral type — mass (in solar units) relation for various luminosity classes. See Table 1 for data.

Errors of estimated L, T and m values were calculated from a derivative of the corresponding relation for a particular spectrum, assuming that spectral types are known with one subclass accuracy. Special attention was paid to spectral types with worse known temperature class (e.g., B5-8) or luminosity class (e.g., F8IV-V), and errors were calculated in those cases correspondingly. For the (usually worse known) spectrum-mass relation, a width of the relation was also included in error estimation.

In a case of absence of spectral classification, components effective temperature ratio can also be estimated from observed depths of light curve minima A_1, A_2, if limb

TABLE 1. Spectrum-mass relations

Spectral type	Luminosity class, LC I	II	III	IV	V
O5	30				51.60
O6			28.0		
O7					27.80
O8	26		24.0		
O9	20	28.0		27.0	
B0	14	23.0	21.0		15.00
B1				12.0	
B2		18.0	16.3	7.1	
B4				3.0	
B5	10	13.0	8.0		4.95
B6				2.2	
B8			5.3		
A0		11.0	4.3		2.40
A2			3.8		
A4				1.9	
A5		8.2	3.3		1.88
A8			2.5		
F0		7.1			1.46
F2			1.9	1.7	
F5		6.8		1.6	1.27
G0		6.9	1.7	1.3	1.10
G4	3		2.8		
G5		7.3		1.3	1.02
G8			2.6	1.5	
G9	6				
K0		7.8		1.5	0.88
K2			2.0	1.3	
K3				1.0	
K4	9				
K5		8.4			0.65
M0	18	9.6			0.52
M2	20				
M5					0.18
M6					0.10
M7					0.04

darkening is neglected:

$$\frac{T_2}{T_1} = \sqrt[4]{\frac{J_2}{J_1}} = \sqrt[4]{\frac{1+0.4A_2}{1+0.4A_1}}, \tag{5}$$

where J_i is surface brightness of i-th component. In particular, $T_1 = T_2$ when $dA \equiv A_1 - A_2 = 0$. Some 392 systems in CEV catalogue have $dA = 0$ and other 628 systems have $dA \leq \sigma$. Among these 1020 systems, 111 systems have $A_1 = A_2 = 2.5\log 2 \pm \sigma$, which means that the two components are identical, i.e. $T_1 = T_2$ and $R_1 = R_2$. He we accept σ to be $0^m.05$.

Also, as it was shown in Malkov et al. [2007], when radii of the components are equal:

$$10^{-0.4A_1} + 10^{-0.4A_2} = 1, \tag{6}$$

and some 10 detached systems in CEV satisfy this criterion.

In a case of *total eclipse*, some additional relations can be drawn for light curve parameters and characteristics of components. In particular, radius ratio and luminosity ratio can be estimated from depths of minima A_1 and A_2. Let L_i be luminosity of i-th component, and $L_{max} \equiv L_1 + L_2$, L_{min1} and L_{min2} be system luminosity in maximum, in primary minimum and in secondary minimum, respectively.

If the primary (i.e., hotter) component is larger, than

$$L_{min2} = L_1, \qquad L_{min1} = (1 - (R_2/R_1)^2)L_1 + L_2. \tag{7}$$

Consequently,

$$\frac{R_2}{R_1} = \sqrt{\frac{L_{max} - L_{min1}}{L_{min2}}} = \sqrt{10^{0.4A_2} - 10^{-0.4dA}}, \tag{8}$$

and

$$\frac{L_2}{L_1} = \frac{L_{max}}{L_{min2}} - 1 = 10^{0.4A_2} - 1. \tag{9}$$

If the primary component is smaller, than

$$L_{min1} = L_2, \qquad L_{min2} = L_1 + (1 - (R_1/R_2)^2)L_2. \tag{10}$$

Consequently,

$$\frac{R_1}{R_2} = \sqrt{\frac{L_{max} - L_{min2}}{L_{min1}}} = \sqrt{10^{0.4A_1} - 10^{0.4dA}}, \tag{11}$$

and

$$\frac{L_1}{L_2} = \frac{L_{max}}{L_{min1}} - 1 = 10^{0.4A_1} - 1. \tag{12}$$

Altogether 493 systems in CEV catalogue are indicated to have total eclipse,

3.2. Input data

In this section we describe the process of compilation of an input list of eclipsing binaries, for their consequent evaluation with the procedure, described in Section 3.1. We have selected eclipsing systems from the Catalogue of Eclipsing Variables, CEV, Malkov et al. [2006].

The following parameters should be available for a binary to put to the parameterization procedure: period, primary spectrum, relative radii of both components, evolutionary class.

Period. Period is always drawn from CEV.

Evolutionary class. System evolutionary class, when unknown from original study, is estimated in accordance with the procedure, described in Malkov et al. [2007], [2008] and included in CEV.

In a case when evolutionary stage of binary remains unknown from CEV classifier, it can be assigned according to the shape of the light curve (given in CEV). Our assumptions were based on results of Giuricin et al. [1983]. According to that study, EW-type light curve is a good indicator of a contact system (though not conversely). Regarding EB-type binaries, we have adopted the contact stage if period $P \leq 0.^{d}6$; for longer periods semi-detached (O and B spectra) and detached (later spectra) stages were assigned. Also, EA-type light curve serves as a good indicator of semi-detached nature of a system if depths of minima satisfy $A_1/A_2 > 0.4$ and $A_1 > 1$.

Relative radius. Relative radii of the components are not given in CEV. The sources of r_1, r_2 are:

- Catalogues of detached main-sequence systems by Svechnikov & Perevozkina [1999] and Perevozkina & Svechnikov [1999].
- Lists of semi-detached binaries by Surkova & Svechnikov [2004].
- Budding et al. [2004] catalogue of semi-detached systems.
- Dryomova et al. [2005] catalogue of near-contact systems.
- Bondarenko & Perevozkina [1996] catalogue of early-type contact systems.
- List of contact binary stars by Csizmadia & Klagyivik [2004].
- A list of observed data for some 1000 eclipsing binaries (Dworak 1980, unpublished) that was prepared for the compilation of Brancewicz & Dworak [1980] catalogue.

Relative radius estimated values with weight 6 to 9 from Catalogue of approximate photometric and absolute elements of eclipsing variable stars by Svechnikov & Kuznetsova [1990] were also used. As an analysis shows, they differ from observational ones by less than 10%. However, it does not apply to late-type contact systems, and for that latter type a list of Lyubimova (unpublished) can be used.

When relative radii was not available, but phase of totality was observed, radius estimation was made according to

$$r_l + r_s = \frac{\pi D}{\sin i}, \qquad r_l - r_s = \frac{\pi d}{\sin i}, \tag{13}$$

and, consequently,

$$r_s = \frac{\pi}{2}\frac{D-d}{\sin i}, \qquad r_l = \frac{\pi}{2}\frac{D+d}{\sin i}, \tag{14}$$

where r_s, r_l are radii of smaller and larger components, correspondingly, D is duration of minimum in decimals of the period, d is duration of the phase of constant light at the bottom in decimals of the period, i is inclination. Totality duration values are given in the Wood et al. [1980] catalogue and in the Catalogue of Chromospherically Active

Binary Stars by Strassmeier et al. [1993]. Inclination i was assumed to be 90° in all such cases, when unknown. In the case of total eclipse r_l and r_s can be estimated from D and d. In the case of partial eclipse, when only D is known, just $r_l + r_s$ can be estimated.

Spectral type. Spectral types of the components were taken from CEV when available. Beside that, almost all catalogues listed in the previous (relative radius) section contain information on system spectral type as well. Also, Polushina [2004] and Maceroni [1996] catalogues served as sources of spectral classification of eclipsing binaries. Spectral type of binaries can be estimated also from multicolor photometry, when available.

Beside period, primary spectrum, relative radii of both components, and evolutionary class we also supply selected stars with other (optional) parameters, when available in CEV: secondary spectrum and depth of minima.

3.3. Detached systems

Systems with unevolved and independently evolved components. This type of detached eclipsing binaries consists of the following groups:

- main sequence systems (marked as DM in CEV);
- binaries, which are thought to be in a pre-main sequence evolutionary stage (e.g., BM Ori): few such systems are known, and in catalogues they are usually included in the previous group;
- systems with at least one supergiant or late type giant (marked as DG in CEV, e.g., VV Cep).

The basic procedure for determination of fundamental parameters of a binary system, consisting of two unevolved (or evolved separately) components, with known period, relative radii of the components and two-dimensional spectral classification of the primary can be written as follows, depending on available data about secondary spectrum:

If *two-dimensional spectral classification for secondary is known*:

- L, T, m were determined for both components from equations (4), with corresponding errors $\sigma L, \sigma T, \sigma m$.
- R was calculated for each component from equation (1).
- a values are calculated from equation (3) for each component and also from equation (2).
- Three values of a calculated in the previous item were compared, and a was varied in that range with minimization of the function

$$D = \sqrt{\sum_{i=1}^{2} \left(\frac{m_i - m_i^s}{\sigma m_i}\right)^2 + \left(\frac{L_i - L_i^s}{\sigma L_i}\right)^2}, \tag{15}$$

where m, L of the two components are adjustable parameters. Their spectrum based values (determined in the first item of the procedure) are designated with upper index "s". Temperatures of the components are considered to be fixed parameters.

If only *one-dimensional spectral classification for secondary is known* than the procedure described above was repeated for each luminosity class. The luminosity class with a minimum D value from equation (15) was assigned to the secondary, together with corresponding m, R, L, T values.

The following procedure was applied to eclipsing binaries with known two-dimensional spectral classification of primary and *unknown spectral classification of secondary* to determine its fundamental parameters.

- L, T, m were determined for primary from equations (4), with corresponding errors $\sigma L, \sigma T, \sigma m$.
- Consequently, primary absolute radius R_1 and system semi-major axis a were calculated from equations (1) and (3), respectively.
- Consequently, secondary mass m_2 and absolute radius R_2 were calculated from equations (2) and (3), respectively.
- If observed depths of both minima A_1, A_2 were available, the secondary effective temperature T was estimated from equation (5). Then the secondary luminosity L_2 was calculated from (1).

Even if the primary luminosity class is not available (i.e., only *one-dimensional spectral classification for primary is known*), the procedure can be applied to some binary systems. In some such cases a two-dimensional spectral classification for secondary is available. The procedure "two-dimensional primary spectrum + one-dimensional secondary spectrum", described above, was used there (with secondary $\leftrightarrow$ primary mutual substitution). In other cases we assumed primary to be a giant if its spectral type is later than G6, and to be a main sequence star otherwise. Then one of the procedures described above was applied, depending on availability of secondary spectrum.

The procedure described above is based essentially on equations (1)-(4). However, an alternative way to parameterize an eclipsing binary system can be proposed. One can use a stellar evolution model to connect observational (radius R, effective temperature T) stellar parameters with fundamental (mass m, age τ) ones:

$$T_i = f_T(m_i, \tau), \quad R_i = f_R(m_i, \tau), \quad i = 1,2 \tag{16}$$

Here we assume that the components are coeval. Primary temperature can also be estimated from spectral type

$$T_1 = T(Sp_1) \tag{17}$$

Together with Eqs. (2, 3) one has 8 equations with 8 unknowns $(m_1, m_2, R_1, R_2, T_1, T_2, a, \tau)$. If a secondary spectral type is known, an extra equation can be added in the equation system, similar to Eq. (17):

$$T_2 = T(Sp_2). \tag{18}$$

For single star evolution case, analytic formulae for MS and Hertzsprung gap stages for stars with masses from 0.5 to 50 solar mass from Hurley et al. [2000] can be used as eqs. (16).

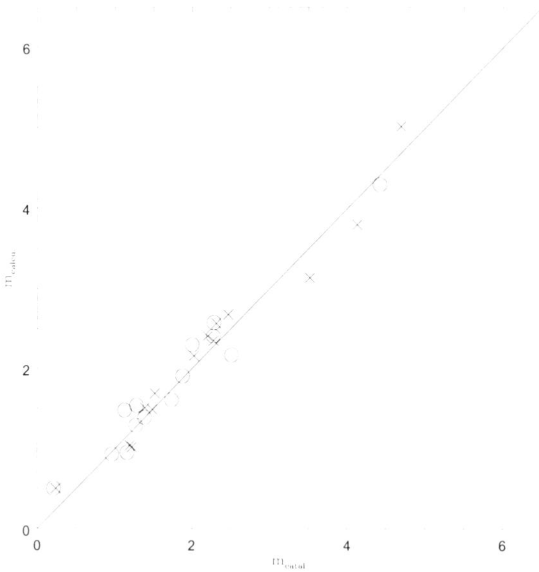

FIGURE 2. Catalogued (dynamical) vs. calculated mass, in solar units, of primaries (crosses) and secondaries (circles) components of detached main-sequence systems. Errors for catalogued data do not exceed symbol size, errors for calculated data are about 20%.

To check the procedure, detached main-sequence systems with known dynamical masses from Svechnikov & Perevozkina [1999] were selected. Among 114 catalogued systems, 14 have spectroscopic masses and only primary spectral classification (we do not consider systems with both components spectral type). The result of comparison of estimated masses with empirical ones is shown in Fig. 2. As it turned out, for late B-stars Popper's [1980] "spectral type — effective temperature" scale better satisfies observations than de Jager & Nieuwenhuijzen's [1987] one.

Systems with two subgiants (AR Lac systems, RS CVn systems). RS CVn-type binaries (marked as RS in GCVS and as DR in CEV) typically are expected to have negligible mass exchange (though the majority of known RS CVn binaries are classified as detached in GCVS). In spite of the fact that some, comparatively small, number of detached binaries, which are likely have undergone (or to be undergoing) mass exchange with possible mass ratio reversal between the two components, presents in the sample, we nevertheless applied the basic procedure to all detached binaries.

3.4. Semi-detached systems

Classical algols. Classical algols are the semi-detached systems, in which the more massive component lies in the range from middle B to early F, and the other is of type F or later. Two reasonable assumptions are usually made to derive the absolute elements of such systems:

- The primary (hotter, brighter and more massive) component is assumed to have a normal mass for its spectral type, to be not overluminous compared to MS stars in the mass-luminosity plane (Giuricin et al. [1983]), and can be considered as a main sequence star. Note that the primary can have larger (if it is a donor) or smaller (if it is an accretor) radius than the secondary. Statistics of semi-detached systems in Surkova & Svechnikov [2004] and Budding et al. [2004] catalogues gives a ratio of such systems as about 3:7.
- The secondary (cooler, overluminous and oversized) component is assumed to be in contact with its Roche lobe, and, according to Plavec and Kratoshvil [1964], the system mass ratio $q \equiv m_2/m_1$ can be determined from the secondary relative radius r_2 (for $q \geq 0.1$):
$$\log q = 3.6456 \log r_2 + 1.5443. \tag{19}$$
The main error source of q determined in such a way is an uncertainty of r_2: the functional dependence of the Roche lobe radius on the mass ratio is such that small changes in the chosen radius result in large changes of the derived mass ratio. Assuming the typical accuracy of cataloged r_2 values to be 0.01 we have estimated an average error of q determination to be about 0.05.

The first assumption allows us to determine the primary mass, luminosity and temperature from its spectrum. Consequently, the following procedure was used:

- L, T, m were determined for primary from equations (4), with corresponding errors $\sigma L, \sigma T, \sigma m$. In a case of one-dimensional primary spectral classification, LC=V was assumed.
- Primary R was calculated from equation (1). Then a was calculated from equation (3).
- Mass ratio q was calculated from equation (19). Then a was calculated from equation (2) (using m_1 estimated above and $m_2 = qm_1$).
- Two values of a calculated in the two previous items were compared. Variations of m_2 in this range were performed to minimize the function
$$D = \sqrt{\left(\frac{m_1 - m_1^s}{\sigma m_1}\right)^2 + \left(\frac{L_1 - L_1^s}{\sigma L_1}\right)^2 + \left(\frac{q - q^r}{\sigma q}\right)^2} \tag{20}$$
Here m_1, L_1, q are adjustable parameters, m_1^s, L_1^s are spectrum based values (determined in the first item of the procedure), and q^r is r_2-based value determined from equation (19).
- Secondary R was calculated from equation (3).

- The secondary spectrum, when available, was served as a source for T_2 estimation and consequent L_2 calculation from equation (1). Stellar effective temperature depends weakly on the luminosity class, that is why, as an approximation, we could use a subgiant (the most suitable Algol-type secondary stage) tabular T value in a case of one-dimensional spectral classification. Else if observed depths of both minima were available, the secondary effective temperature was estimated from equation (5).

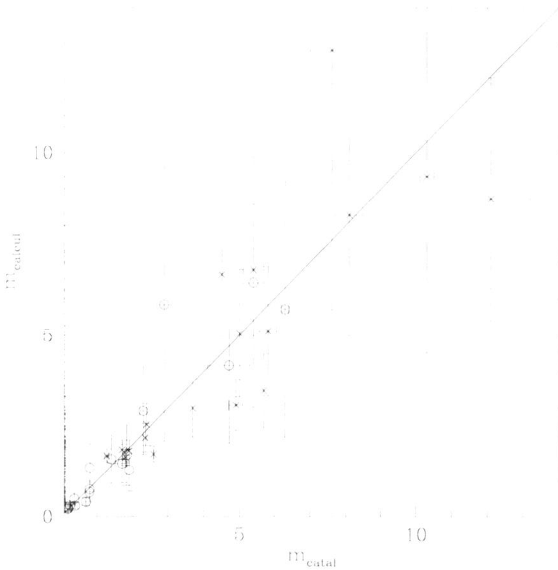

FIGURE 3. Catalogued (dynamical) vs. calculated mass, in solar units, of primaries (crosses) and secondaries (circles) components of semi-detached systems, with their errors.

To check the procedure, 18 detached systems with known dynamical masses and radii from Surkova & Svechnikov [2004] were selected. The result of comparison of estimated masses and radii with empirical ones is shown in Figs. 3 and 4.

Hot and cool semi-detached systems. Popper [1980] in his comprehensive review on stellar masses distinguishes (somewhat arbitrarily), besides typical algols, two additional groups: *hot semi-detached systems* (in which the hotter component is an early B-type star and the cooler of type B or early A) and *cool semi-detached systems*, with the later subgiant and giant. The two groups consist of four systems each. Nelson & Eggleton [2001] in their theoretical study also discuss hot and cool semi-detached systems.

Double-contact (β Lyr) systems. According to Wilson [1979], both components in such a binary fill their Roche lobes exactly but do not touch each other due to asynchronous rotation of at least one of them. Svechnikov [1969] considers β Lyr to be "on its way to a detached system with a subgiant". However, there is a doubt that β Lyr can represent a sub-class, because more recent investigations (see, e.g. Linnel, [2002]) confirm the uniqueness of this system.

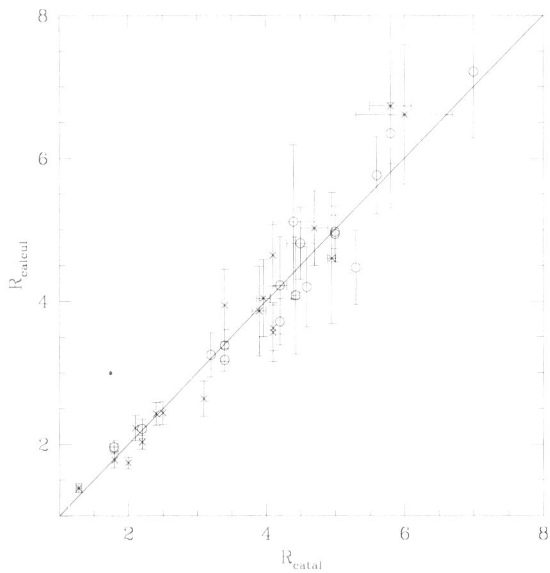

FIGURE 4. Catalogued (dynamical) vs. calculated radius, in solar units, of primaries (crosses) and secondaries (circles) components of semi-detached systems, with their errors. The V356 Sgr primary ($R_{catal} = 14.0 \pm 1.5$, $R_{calcul} = 12.53 \pm 1.85$) is not plotted.

Undermassive (R CMa) systems. Both components in such systems are overluminous, oversized and hotter for their derived masses; neither component can be considered as a main sequence star (see Sarma et al.'s [1996] study of the prototype).

Systems in early stages of mass transfer. In some early-type detached systems the more massive component is evolving off the main sequence and almost filling its Roche lobe thereby starting mass transfer (see, e.g., Kallrath & Strassmeier's [2000] study of BF Aur). Then a very rapid mass transfer takes place from the primary to the secondary, proceeding on a thermal time scale. Examples are TT Her (Milano et al., [1989]) classified as a contact system in the GCVS and AB Cru (Lorenz et al., [1994]) classified as a detached system in the GCVS. The latter has already undergone a reversal of its mass ratio, so mass transfer from the (formerly more massive) secondary towards the primary is observed. In these cases a period change or/and a mass ratio reversal support the assumption that mass exchange has already started or is about to start.

Broken-contact systems. If the energy transfer in a close binary is insufficiently effective, a cycle consists of a long quiet phase in good contact and a short violent phase with rapid changes between contact, semi-detached and detached configurations (Kähler, [2002]). One such system observed in a semi-detached stage is CN And (Van Hamme et al., [2001]). Broken-contact systems exhibit orbital periods in the range $0.3 - 0.7$ days and the spectral type is around F-G.

These six groups consist of a relatively small number of binaries; that is why, in spite of the fact that the equations (4) no longer valid for the primary, we considered these systems statistically insignificant and we did not separate them from other semi-detached pairs (some of stars from the latter two groups are classified as contact or detached binaries in GCVS).

Systems with compact objects. Our procedure can not be applied to systems, containing compact objects: white dwarfs, neutron stars, black holes. Particularly, the procedure can not be applied to X-ray binaries: close systems with a main sequence (or giant, or degenerate) primary filling its Roche lobe and transferring matter to the compact object – a neutron star or a black hole. To remove such systems from the sample, we have cross-checked our list with the following catalogues:

- Catalog of evolved close binary stars (Cherepashchuk et al. [1986]).
- Catalogue of cataclysmic binaries, low-mass X-ray binaries and related objects, sixth edition (Ritter & Kolb [2003]).
- Catalogues of high- (Liu et al. [2006]) and low- (Liu et al. [2007]) mass X-ray binaries.

3.5. Contact systems

According to Wilson's [1979] notation, they are in fact overcontact systems, where each component has a surface larger than its Roche lobe. They are synchronous, circular orbit systems with common envelope.

Late-type systems. Spectrum of primary in such systems (called also W UMa systems) is usually later than about F0.

The following assumptions can be made to estimate fundamental parameters of W UMa systems. Since the systems are in contact the two stars have the same surface potential. This can be expressed by an approximate mass-radius relation (for detailes see, e.g., de Loore & Doom [1992], p. 360):

$$\frac{m_1}{m_2} = \left(\frac{R_1}{R_2}\right)^2 \equiv \left(\frac{r_1}{r_2}\right)^2 \tag{21}$$

Shape of a typical W UMa system light curve indicates that the two components have comparable temperatures. If so, the approximate mass-luminosity relation may be derived from equations (1) and (21):

$$\frac{L_1}{L_2} = \frac{m_1}{m_2} \tag{22}$$

This was generally confirmed by an empirical relation that was derived by Oh [1999] from an analysis of 761 contact binaries in Svechnikov & Kuznetsova [1990] catalogue:

$$\frac{L_1}{L_2} = 0.01 + 0.89\frac{m_1}{m_2} \tag{23}$$

Another useful relation was found by Yang & Liu [2001], who analyzed the nature of the secondary overluminosity in contact systems. It was shown that the overluminosity ΔL_2 of the secondary is clearly related to the mass of the primary m_1 (see their Fig. 4). This relation for late-type systems was roughly approximated by us as

$$m_1 = 0.7(\Delta L_2 + 1) \tag{24}$$

The overluminosity in solar unites is defined as

$$\Delta L_2 = L_2 - m_2^4, \tag{25}$$

where L_2 and m_2 are observed luminosity and mass of the secondary, respectively.

So, the following procedure can be applied to estimate fundamental parameters of late-type contact systems.

- The temperature of components ($T_1 = T_2$) is estimated from the system spectrum (note that in this case even a one-dimensional spectrum serves as a good enough approximation).
- Primary mass $m_1 = 1$ is assigned, as a first approximation (according to observational statistics of late-type contact binaries, the total mass is 0.9 to 2.3). Then an iteration process begins.
- Relative radii serves for mass ratio q calculation from equation (21). Then m_2 is calculated from q and m_1.
- a is calculated from equation (2). Then R, L of the components is calculated from equations (3) and (1), consequently.
- A primary mass is estimated from equation (24). Then the iteration process is repeated from the item 3 of this procedure untill the m_1 change becomes negligible.

Early-type systems. Such contact systems of early spectra have both components close to their Roche lobes. The spectrum of the more massive component is usually not later than about F0 (according to Svechnikov [1969]) or even A0 (according to Pribulla et al. [2003]).

Equations (21) and (24) are valid for this type of binaries as well. But early-type contact binaries show higher temperature difference, so T_2 can not be estimated from the primary spectrum. It means that the procedure, described in the previous paragraph can be applied only if the secondary temperature can be estimated from the secondary spectrum or from equation (5).

In other cases a period-luminosity relation was used to reduce the number of unknowns to the number of equations. An empirical $m_{ph} - P$ relation was given by Richter [1968]. Using a modern data on early type contact binaries we have found an approximate period-luminosity relation:

$$\log L = 8 \log P + 2 \tag{26}$$

that allows us to apply the following procedure to parameterization of early-type contact systems.

- The temperature of the primary is estimated from the spectrum.

- Primary luminosity is estimated from equation (26). Then primary and secondary radii are consequently calculated from equations (1) and (21).
- a is calculated from equation (3).
- Sum of masses, calculated from equation (2), and their ratio, calculated from equation (21), produces individual masses. Then secondary luminosity and temperature are calculated, consequently, from equations (24) and (1).

Near-contact systems (CB). This class was introduced by Svechnikov [1969] as being similar to W UMa-stars, where both components do not fill their Roche lobes and their physical characteristics are similar to those of W UMa-type systems. They are classified as ~KW in Svechnikov's notation, as DW in GCVS, and as CB in CEV. They are also called short-period RS CVn-type or W UMa-similar non-contact systems. Formally they can be referred to as detached or semi-detached systems, but their light curves are peculiar and distorted by gaseous streams, and the brightness of the components is slightly variable, which makes them similar to contact systems of late spectral type. Pribulla et al. [2003] designate them as B-type contact systems (reminding β Lyr light curve), whose components are in physical but not in thermal contact. These are also sometimes called binaries with two stars of very different effective temperature enclosed in a common envelope with the secondary located below and to the left of the main sequence in the HRD. For such systems we used the procedure, described in Section 3.3 (detached systems), assuming that no secondary spectral type is available.

4. CONCLUSIONS

In the present study we briefly discuss modern catalogues and lists of eclipsing binaries and present a set of procedures for parameterization of their components. To check their usability, some procedures are applied to eclipsing binaries with known dynamical masses.

Our next steps are to develop classification procedures for other, more marginal types of close binaries and to apply all procedures to other lists of eclipsing binaries with known relative radii of components.

Possible astrophysical applications of the described results are very numerous. Classified and parameterized eclipsing systems can serve for study of fine structure of the MLR for intermediate masses and deviations from the MLR caused by evolution, metallicity, rotation and other effects. More accurate information on status of some types of EB (essentially contact and near-contact) in evolutionary scheme and fraction of interactive binaries population on the subgiant (IV) sequence of the HRD can also be obtained. At last, "marginal" systems displaying exceptional values of orbital or/and component parameters can be selected for further study.

ACKNOWLEDGMENTS

I thank Ekaterina Avvakumova, Eduard Oblak, Jordi Torra, Dana Kovaleva, Sergej Sichevskij, Leonid Kalinichenko and Marat Kazanov for collaboration and valuable comments. This work has been supported by Russian Foundation for Fundamental Research grants 09-02-00520 and 10-02-00426, by the Federal Science and Innovations Agency under contract 02.740.11.0247, by the Federal target-oriented program "Scientific and pedagogical staff for innovation Russia" (contract No. P1195), and by the Presidium RAS program "Leading Scientific Schools Support" 4354.2008.2. This research has made use of the SIMBAD database, operated at the Centre de Données astronomiques de Strasbourg, and of NASA's Astrophysics Data System Bibliographic Services. The use of the WDS, GCVS, SB9 and VizieR databases is also acknowledged.

REFERENCES

Belikov A.N. 1995, Bull. Inf. CDS 47, 9

Bondarenko I.I., Perevozkina E.L. 1996, "Catalogue of Photometric, Geometric, and Absolute elements of contact binary stars of the early spectral type". Publ. Odessa Obs. 9, 20

Brancewicz H. K. & Dworak T. Z. 1980, Acta Astron. 30, 502

Budding E., Erdem A., Çiçek C. et al. 2004, A&A 417, 263

Cherepashchuk A., Katysheva N., Khruzina T., Shugarov S. 1996, "Catalog of Evolved Close Binary Stars", Brussel: Gordon and Breach Science Publishers, parts 1 and 2.

Csizmadia Sz., Klagyivik P. 2004 Astron. Astrophys. 426, 1001

Dryomova G. N., Perevozkina E. L. & Svechnikov M. A. 2005, A&A 437, 375

Dworak T.Z. 1975, Acta Astronomica 25, 383

Gettel S.J., Geske M.T., McKay T.A. 2006, Astron. J., 131, 621

Giuricin G., Mardirossian F., Mezzetti M. 1983, A&AS 54, 211

Hurley J.R., Pols O.R., Tout, C.A. 2000, MNRAS 315, 543

de Jager C. & Nieuwenhuijzen H. 1987, A&A 177, 217

Kähler H. 2002, A&A 395, 907

Kaygorodov P., Debray B., Kolesnikov N. et al. 2012, Baltic Astronomy, in press.

Kallrath J. & Strassmeier K. G. 2000, A&A 362, 673

Kraicheva Z., Popova E., Tutukov A., Yungelson L. 1980, Bull. Inf. CDS 19, 71

Linnel A. P. 2002, MNRAS 334, 963

Liu Q.Z., van Paradijs J., van den Heuvel E. P. J. 2006, A&A 455, 1165

Liu Q.Z., van Paradijs J., van den Heuvel E. P. J. 2007, A&A 469, 807L

de Loore C. W. H. & Doom C. 1992, Structure and Evolution of Single and Binary Stars. In Astrophysics and Space Science Library, Kluwer Academic Publishers, vol. 179

Lorenz R., Mayer P., Drechsel H. 1994, A&A 291, 185

Maceroni C., & van't Veer F. 1996, A&A 311, 523

Malkov O. Yu. 1993, Bull. Inf. CDS 42, 27

Malkov O., Oblak E., Snegireva E.A., Torra J. 2006, A&A 446, 785

Malkov O., Oblak E., Avvakumova E.A., Torra J. 2007, A&A, 465, 549

Malkov O., Kalinichenko L., Kazanov M., Oblak E. 2008, in *ADASS XVII Conference*, eds. Bob Argyle, Peter Bunclark, Jim Lewis, London, Sep 2007, ASP Conf. Ser., 394, 381

Milano L., Barone F., Mancuso S., Russo G., Vittone A. A. 1989, A&A 210, 181

Nelson C. A. & Eggleton P. P. 2001, ApJ 552, 664

Oh K.-D. 1999, Journal of Astronomy and Space Sciences 16, 31

Perevozkina E. L. & Svechnikov M. A. 1999, "Catalogue of orbital elements, masses and luminosities of eclipsing variable stars (detached main-sequence type) with known elements of photometric orbit and unknown spectroscopic elements", Ekaterinbourg, Ural University, 1999, p.122

Plavec M. & Kratoshvil P. 1964, Bull. Astron. Inst. Czech. 15, 165

Polushina T.S. 2004, Astron. and Astrophys. Transactions 23, 213

Popper D.M. 1980, ARAA 18, 115

Pribulla T., Kreiner J. M. & Tremko J. 2003, Contributions of the Astronomical Observatory Skalnate Pleso 33, 38

Richter G. 1968, Veröff. Sternw. Sonneberg 7, part 3

Ritter H. & Kolb U. 2003, A&A 404, 301

Samus N.N., Durlevich O.V., Kazarovets E V., Kireeva N.N., Pastukhova E.N., Zharova A.V. et al. General Catalog of Variable Stars (GCVS database, Version 2011 Jan), http://www.sai.msu.su/gcvs/gcvs/iii/html/

Sarma M. B. K., Vivekananda Rao P., Abhyankar K. D. 1996, ApJ 458, 371

Shaw J.S. 1994, MmSAI, 65, 95

Strassmeier K. G., Hall D. S., Fekel F. C., Scheck M. 1993, A&AS 100, 173

Svechnikov M. A. 1969, "Catalogue of orbital elements, masses and luminosities of close binary stars", Sverdlovsk, Publ. Ural University, Ser. Astronomia, issue 5, N 88.

Svechnikov M. A. & Kuznetsova Eh. F. 1990, "Catalogue of approximate photometric and absolute elements of eclipsing variable stars", Vols 1-2, Sverdlovsk, Ural University

Svechnikov M. A. & Perevozkina E. L. 1999, "Catalogue of orbital elements, masses and luminosities of eclipsing variable stars (detached main-sequence type) with known photometric and spectroscopic elements", Ekaterinbourg, Ural University, 1999, p.5

Surkova L.P. & Svechnikov M.A. 2004, "Catalog of Photometric, Geometric and Absolute elements of semi-detached eclipsing binaries with known photometric or spectroscopic orbits", VizieR on-line database: V/115.

Svechnikov M. A. & Taidakova T. A. 1984, SvA 28, 84

Van Hamme W., Samec R. G., Gothard N. W. et al. 2001, AJ 122, 3436

Wilson R. E. 1979, ApJ 234, 1054

Wood F. B., Oliver J. P., Florkowski D. R., Koch R. H. 1980, Publ. Dept. of Astron., Univ. of Florida, Vol. I

Yang Y. & Liu Q. 2001, AJ 122, 425

Emission Objects in CYG OB7. Flare Activity Variation of UV CETI.

N.D.Melikian[a], V.S.Tamazian[b], A.A.Karapetian[a], A.L.Samsonyan[a]

[a] *V.A.Ambartsumian Byurakan Astrophysical Observatory, Armenia; e-mail: nmelikia@bao.sci.am*
[b] *Ramón María Aller Astronomical Observatory, University of Santiago de Compostela, Spain; e-mail: vakhtang.tamazian@usc.es*

Keywords: stellar activity, binary stars
PACS: 97.10.Jb; **97.21.+a**; 97.30.Nr; 97.80.Di

1. INTRODUCTION

T Tauri and UV Ceti type stars, as well as Herbig-Haro (HH) objects represent the youngest low-luminosity objects in stellar associations. Close relation in evolutionary status makes their study extremely important in order to clarify the evolutionary sequence of low-mass stellar objects. It is worth noting that overwhelming majority of these objects was discovered during different Hα surveys and photographic monitoring.

One of the Hα surveys has been performed with the 1m Schmidt telescope of Byurakan Astrophysical Observatory (BAO), equipped with a 4° objective prism. More than 200 new Hα objects have been discovered in 6 selected dark cloud regions with a total area of about 100sq.degrees [1-5]. This survey carried out during 10 years(1979-1989) showed that more than 10% of discovered stars with Hα emission exhibit spectral line intensity changes both on short (1 – 3 months) and long (1 -10 years) timescales.

New observations aimed to a more detailed study of a small 6'x11' area around two previously discovered emission-line stars[5] were carried out using the 2.6m telescope of BAO. They allow to detect for the first time 3 HH objects, about 50 Hα emission line stars, 2 cometary nebulae, 2 UV Ceti type flare stars and one infrared nebula with an interesting structure [6-12] in an area of ~0.3sq.degree. This area is a scrap of the association Cyg OB7 and is being projected onto the dark nebula Khavtassi 141[13] located almost in the middle of the association Cyg OB7 and whose estimated distance is about 400pc[14]. Notice that on the basis of a bright star survey of Hiltner[15] a distance of 740 pc for association Cyg OB7 were derived by Schmidt[16]. Almost all discovered objects are projected on the background of dark nebula. Some of the observed peculiar objects constitute a Trapezium type physical systems [7,8].

As a part of collaboration between astronomers of BAO and Astronomical Observatory R. M. Aller of the University of Santiago de Compostela (Spain), an effort is under way to study relationships between components orbital postion and their physical (flaring) activity in binary non-stationary systems such as young T Tauri

International Workshop: Binaries Inside and Outside the Local Interstellar Bubble
AIP Conf. Proc. 1452, 73-79 (2012); doi: 10.1063/1.4718588

and UV Ceti type flare stars. Our analysis of the CR Dra system did not reveal a perceptible correlation between flare activity and the distance between the components [17], while such dependence had been noticed in the binary flare FL Vir = Wolf 424 [18]. On the other hand, study of flaring frequency in the binary system UV Ceti allows to suspect its cyclical character with a period of 10-15 years similar to that of the Sun with a period of 22 years [19].

In this report our main results of Cyg OB7 observations with 2.6m telescope of BAO during last 10 years as well as a study of the UV Cet flare activity along its orbital period are presented.

2. OBSERVATIONS

Observations were carried out in 2000-2010 with ByuFOSC and SCORPIO spectral cameras attached to the 2.6m telescope of BAO. These cameras in combination with different filters and grisms are working in several modes: direct imaging in B, V, R, I, Hα and SII filters, the slitless spectra obtaining of field objects with Hα and SII interference filters and long slit spectra of selected objects. The field size with ByuFOSC and SCORPIO is 6' x 11' and – 14' x 14' respectively. Details regarding the observation method, instruments and data reduction procedure are described earlier in [10].

To study flare activity dependence from orbital period of the well known flare star UV Ceti, the observational data from literature have been used. We have chosen the binary system UV Ceti because there is an exceptionally large observational data set, and its orbit has been calculated reliably [20]. At the periastron, the components of this system are at a distance of less than 2 a.u., which is roughly the limit for interaction between magnetic fields of the components in close binary systems [17].

3. EMISSION STARS IN CYG OB 7

During last 10 years an area of ~0.3sq.degree is covered in Cyg OB7 region. More than 50 new emission objects have been discovered in this area, among which 3 groups of new HH objects, 2 new cometary nebulae, 2 UV Ceti type flare stars and a new small nebula of unknown nature. Observed region in Cyg OB 7 is shown in Fig.1, where all newly detected emission objects are shown.

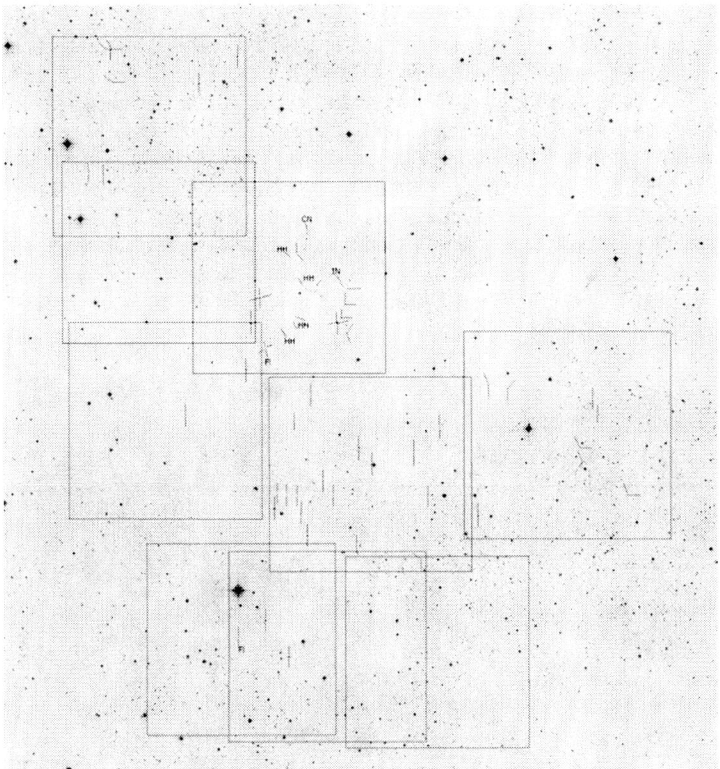

Fig. 1. The observed regions in Cyg OB7

In Fig. 2a, b, c a small field of the region, which includes the nebula of unknown nature is presented. Fig. 2a, is taken from DSS2, and Figures 2b and 2c present our observations obtained in August 2000 and September 2001 in near infrared.

a b c

Fig. 2. The anonymous nebula images from DSS2 (a) and infrared obtained in August 2000 (b) and September 2001 (c)

As it seen in Fig.2 this nebula is not detected on DSS2 images and was absent also on 2MASS survey K band image, obtained in 1999. It is detected for the first time in our observations on August 8 of 2000 only in near infrared [6]. In September 2001 we detect a strong change of its morphology and brightness ($\Delta m \sim 4^m$). Observations with 2.6m telescope of BAO in 2003, 2004, 2005, 2007, 2009 and 2010 show almost stable morphology and brightness of this nebula. The study of spectra of the brightest part of this nebula testifies on existence of embedded in the nebula object.

As it has been noted above, 2 new UV Ceti type flare stars were discovered in this region for the first time. These stars have been detected during observations in the slitless spectroscopy mode. In Figures 3 and 4 the images of fields, were these stars detected, are presented.

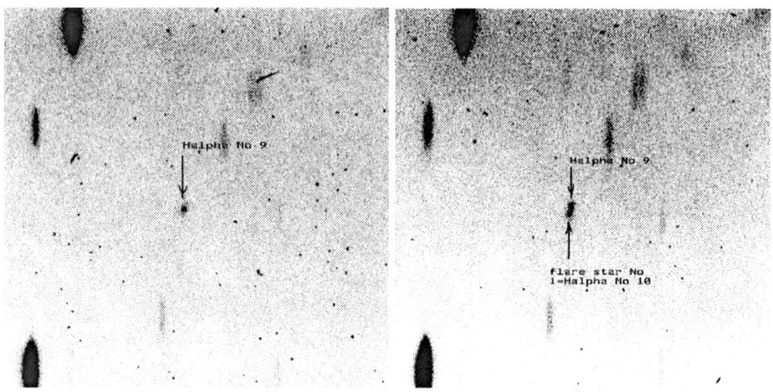

a b

Fig. 3a,b. Identification charts for flare star No 1.

As one can see in Fig.3 flare star No 1 is detected only on image (b) at 1-2 arcsec. from the known emission star No 9.

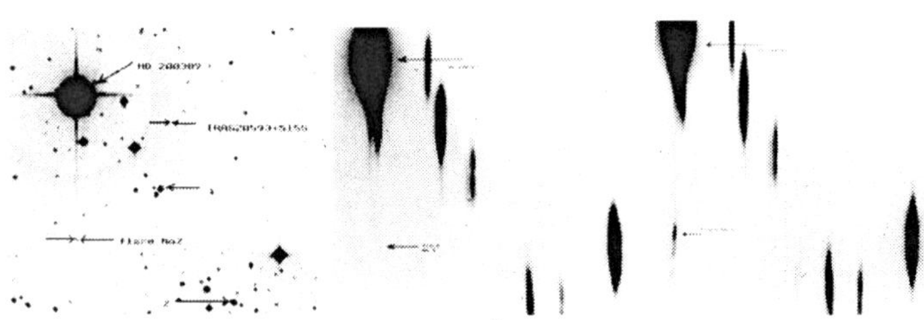

a b c

Fig. 4. DSS2 image of the flare star No. 2 (a) and its slitless spectra before (b) and during the flare (c).

Fig.4a,b,c represents the field of flare star No 2 from DSS2(4a) and slitless spectra of the field objects before(4b) and during the flare up(4c). Flare No 2 is detected in a number of observations, which give possibility to construct the light curve of this flare and to estimate amplitude of flare. The light curve of this flare is shown on Fig. 5.

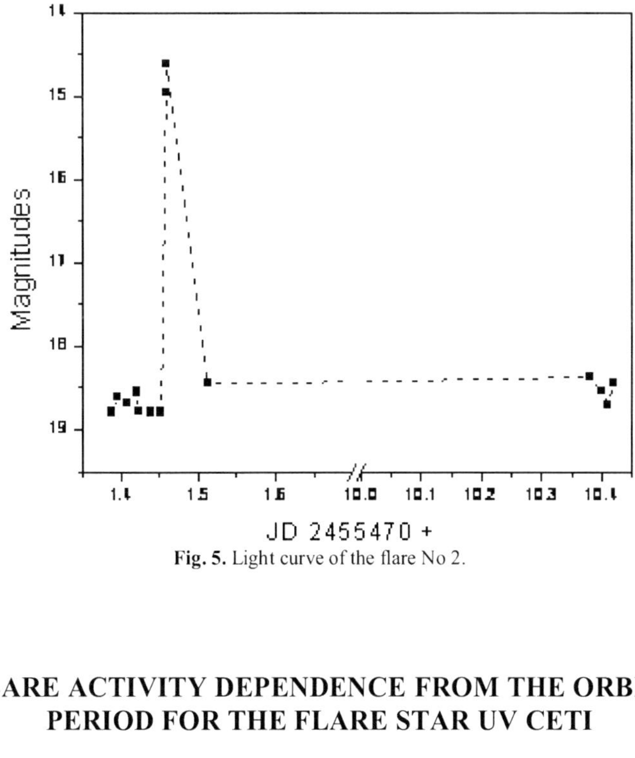

Fig. 5. Light curve of the flare No 2.

4. FLARE ACTIVITY DEPENDENCE FROM THE ORBITAL PERIOD FOR THE FLARE STAR UV CETI

As part of a collaboration between astronomers of BAO and AORMA an effort is under way to determine the orbits of binary stars and study the dependence of their physical activity on the mutual position of the components of binary non-stationary stars, mainly among T Tau stars and UV Ceti type flare stars. An analysis of the data from one of the systems we have chosen (CR Dra) did not reveal a perceptible correlation between flare activity and the distance between the components [17], while a dependence of this sort had been noticed in the binary flare star FL Vir = Wolf 424 [18].

Since 1965 and up to earlier 1980s there were ten international cooperative projects in an extremely active study of flares in stars near the Sun. These observations were made on different telescopes, mainly with electrophotometers. We did not make use of some photographic observations whose errors differed greatly from photoelectric measurements. Even with the photoelectric observations, the use of photomultipliers with different sensitivities leads to different estimates of the flaring frequency, so that, for a given star, different observers often obtained different frequencies within the same observational period. This leads to increased scatter in the flare frequency with time, since observational data are fairly non-uniform. Hence, these data were used after applying to their distribution median filtration which easily eliminates random fluctuations.

As a measure of flaring activity we make use of the quantity N/T (flaring frequency), where N is the number of detected flares and T is the total effective time over which they have been detected.

Fig. 6 shows the variation in the distance between components of UV Ceti (6a), and its flaring frequency change (6b) within the same period. Our results show that flaring frequency clearly depends on the distance between components. It increases rapidly to reach a maximum right at the periastron, while the minimum occurs near the apoastron. This shows that the flaring activity of the star is directly related to its orbital position.

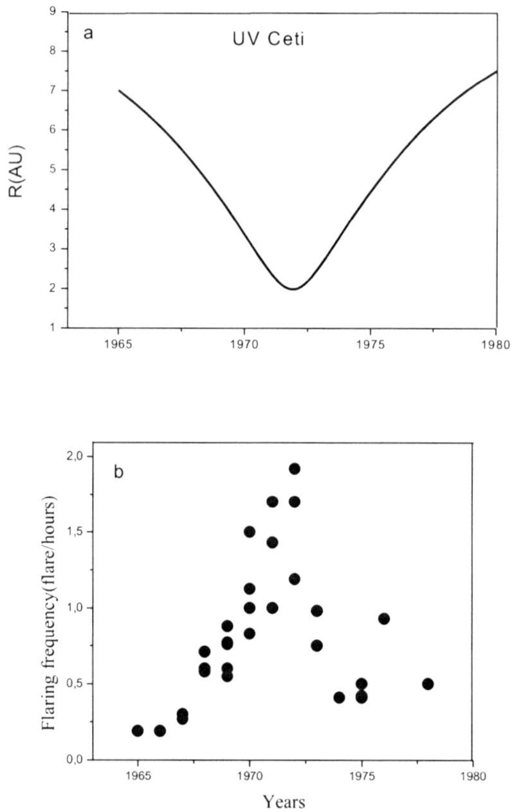

Fig. 6. Orbital position (a) and flaring frequency (b) of UV Cet

Therefore, our results for UV Ceti, are in qualitative agreement with those for FL Vir. If flaring frequency depends on the orbital position of the components so reaching its maximum at periastron and minimum at apoastron, then it is natural to assume that its variation is mainly caused by the change in gravitation force between components and/or an interaction of magnetic fields propagated in the coronae and extra-coronal regions. It is noteworthy that UV Ceti gravitational force at periastron is 18 times greater than at apoastron (about 3.5 times for FL Vir and 2 times for CR Dra). Therefore, one can understand why this dependence shows up very strongly in UV Ceti ,while it is practically unnoticeable in CR Dra. If relationship between flaring frequency and orbital position of components in flare binaries is the same for all such systems, then, it would be possible to solve the inverse problem of determining the orbital period by studying the distribution of the flaring frequency when sufficient observational data sets are available.

REFERENCES

1. N.D.Melikian, V.S.Shevchenko, S.Yu.Melnikov, IBVS, No. 3037-3038 (1987)
2. N.D.Melikian, V.S.Shevchenko, Astrophysics, 32, 169-180 (1990)
3. N.D.Melikian, Astrophysics, 37, 219-228 (1994)
4. N.D.Melikian, A.A.Karapetian, Astrophysics, 39, 27-32 (1996)
5. N.D.Melikian, A.A.Karapetian, L.G.Hakhverdian, A.Ts. Karapetian, Astrophysics, 39, 115-120 (1996)
6. N.D.Melikian, A.A.Karapetian, Astrophysics, 44, 216-221 (2001)
7. H.A.Harutyunian, N.D.Melikian, Astrophysics, 44, 352-358 (2001)
8. N.D.Melikian, A.A.Karapetian, Astrophysics, 46, 282-290 (2003)
9. N.D.Melikian, V.S.Tamazian, A.A.Karapetian, Astrophysics, 49, 320-332 (2006)
10. N.D.Melikian, V.S.Tamazian, A.A.Karapetian, S.K.Balayan, Astrophysics, 50, 10-17 (2007)
11. N.D.Melikian, V.S.Tamazian, J.A.Docobo, A.A.Karapetian, Astrophysics,51, 189-196 (2008)
12. N.D.Melikian, A.A.Karapetian, Astrophysics, 53, 490-496 (2010)
13. D.S.Khavtassi, The Atlas of Galactic Dark Clouds, Abastumani Astrophys.Obs., Georgia, 1960.
14. P.T. de Zeeuw et al., AJ, 117, 354-399 (1999)
15. W.A.Hiltner, ApJS, 2, 389-456 (1956)
16. K.H.Schmidt, Astron.Nachr., 284, 76-78 (1958)
17. V. S. Tamazian, J. A. Docobo, Y. Y. Balega, et al., ApJ. 136, 974-979 (2008)
18. B. R. Pettersen, MNRAS. 368, 1392-1394 (2006)
19. B. R. Pettersen, S. R. Sundland, and S. L. Hawley, ASPC 9, 177-189 (1990)
20. D. Geyer, R. Harrington, and C. Worley, AJ, 95, 1841-1842 (1988)

Plotting the Future of the WDS

William I. Hartkopf

United States Naval Observatory
3450 Massachusetts Ave., NW, Washington, DC 20392-5420

Abstract.
The Washington Double Star Catalog (WDS) and its predecessors have served the double star community for over a century; the current catalog now includes over 800,000 measures of nearly 115,000 pairs and is updated on a nightly basis.

Large interferometric and CCD surveys, as well as recent catalogs such as 2MASS, SDSS, UCAC, etc. have provided much new data for the WDS, but have also highlighted limitations to the current format of the catalog. Multi-wavelength photometry is limited due to a lack of space for filter information. Arcminute–precise WDS designations are proving inadequate to handle binary discoveries in dense fields. Designations for complex hierarchical systems are also proving to be a problem. The Durchmusterung catalogs used for over a century as the primary cross-reference are inadequate for fainter pairs. Astrometric and occultation pairs are not included in the catalog at all. These and other problems will only be exacerbated by massive future surveys such as Gaia.

Some of these issues are illustrated and a few possible solutions presented. It is hoped that discussions can be initiated with users of all the USNO double star catalogs concerning ways to expand the information contained in them, in order to increase their usefulness to the entire double star community.

Keywords: binaries: catalogs
PACS: 97

A BRIEF HISTORY OF THE WDS

Any discussion of the future of the Washington Double Star Catalog must begin with its past. The WDS can trace its lineage back to Sherburn W. Burnham's magnum opus, his *A General Catalogue of Double Stars within 121° of the North Pole* (1906), usually referred to as the "Burnham Double Star Catalog", or BDS. The BDS was the result of over 30 years' work by Burnham, compiling double star observations scattered throughout the literature and making his own followup observations of many of these pairs. After the catalog's publication, Burnham continued to maintain his database, with the intention of publishing a second edition. In 1912, he gave all his data to Eric Doolittle of Flower Observatory, who further updated the database, but died in 1920 before that new edition could be published. The materials were then passed along to Robert Grant Aitken at Lick Observatory, who in 1932 published his *New General Catalogue of Double Stars* (a.k.a., the "Aitken Double Star Catalog", or ADS).

During this same period, Robert Innes (Union Observatory, South Africa) compiled his own catalog of pairs south of $-19°$ (the *Southern Double Star Catalogue*, or SDS). In the 1950's these materials were also sent to Lick Observatory, where they were

International Workshop: Binaries Inside and Outside the Local Interstellar Bubble
AIP Conf. Proc. 1452, 80-86 (2012); doi: 10.1063/1.4718589

merged by Hamilton Jeffers, Willem van den Bos, and Frances Greeby to create the *Index Catalogue of Visual Double Stars, 1961.0* (or IDS).

That same year a young Lick observer named Charles Worley moved to the U.S. Naval Observatory, and a few years later he arranged that all the IDS materials (some 179,000 computer punch cards) be sent to the USNO as well. A printed edition of the new *Washington Double Star Catalog* was published in 1984 by Worley and Geoff Douglass, with a later edition in 1996 ([1]). After Worley's death, Brian Mason, Gary Wycoff, and the author took over maintainance of the database (c.f., [2]; all USNO double star catalogs are now maintained by Mason and Hartkopf). Figure 1 illustrates the considerable growth of the WDS database from 1906 to the present.

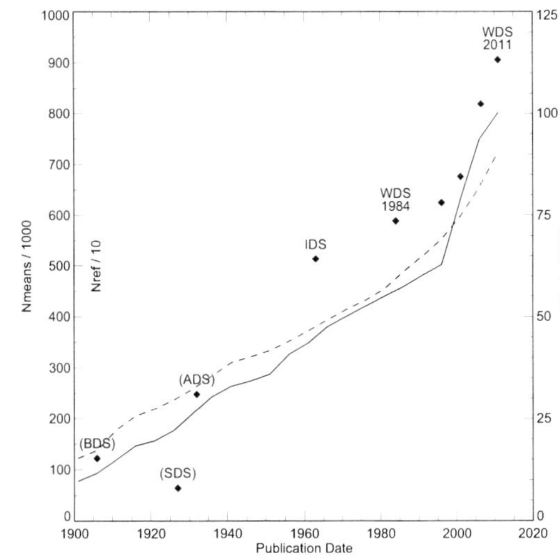

FIGURE 1. Growth of the WDS database from 1906 to the present. Diamonds (right axis) indicate the number of pairs in the database as different editions were published (the earliest catalogs are bracketed with parentheses, as they were not all-sky). WDS 1996, 2001, and 2006 editions are plotted but not labelled. The solid line indicates the number of mean measures, the dashed line the number of publications from which data were extracted (left axis).

WDS: A VICTIM OF ITS OWN SUCCESS?

Much has changed during the past century! Duplicity surveys have expanded greatly in scope, reaching much fainter magnitudes. High resolution surveys, using speckle interferometry, adaptive optics, and now long-baseline interferometry, have discovered many close pairs hitherto unresolvable. Infrared techniques, from IR speckle to 2MASS and other large surveys, have also seen considerable success. As a result, we find ourselves running into a number of limitations due to the current format of the WDS.

We have purposefully been slow to make changes, as we realize that the current catalog is used by many different groups (e.g., the Strasbourg Data Center) and that any changes will force others to modify existing programs. Still, some changes are needed,

WDS	DD	COMP	DATES		N	θ		ρ		MAGS		SPEC	PM		DM	
02425+1045	CHR 200		1994	1994	1	247	247	0.0	0.0	6.3	6.5	A0	-024-024		+10	360
02572-2458	BEU 4	Ca,Cb	2000	2008	3	170	279	0.1	0.0	7.84	.	K2	+015-033			
03082+4057	CSI 1	Aa1,2	2006	2006	12	226	75	0.0	0.0	2.12	.	B8V	+004-001			
03082+4057	LAB 2	Aa,Ab	1973	2008	57	123	307	0.1	0.1	2.12	4.6	B8V	+004-001		+40	673
03082+4057	BU 526	AB	1878	2009	10	155	156	58.8	58.5	2.0	12.7	B8V	+004-001		+40	673
03082+4057	BU 526	AC	1878	2009	10	145	146	67.7	67.7	2.0	12.5	B8V	+004-001			
03083+3250	LDS3456		1960	1960	1	290	290	4.0	4.0	13.88	14.08		+189+008	+189+008		
03084-7033	HLN 19		1892	1998	5	63	66	8.4	8.0	12.55	12.81					
03132+1710	BPM 71		1996	1999	2	169	169	107.2	107.1	11.49	14.38		-015-020	-016-019		
04084-4813	BVD 43		1897	1999	7	237	238	29.8	29.9	9.32	11.65	K2III- G0	-008-034	-007-036		
04331+2410	GHE 13	Aa,Ab	1990	2005	29	330	310	0.4	0.3	11.20	12.1	M				
04331+2410	GHE 13	Aa,b1	1999	2002	7	326	320	0.3	0.3	12.36	14.6					
04331+2410	GHE 13	Aa,b2	1999	2005	21	321	312	0.3	0.2	12.36	15.2					
04331+2410	GHE 13	Ab1,2	1999	2005	22	229	124	0.0	0.0	14.6	15.2					
04332+6945	TDS2892		1991	1991	1	344	344	0.4	0.4	11.29	11.34		-002-023	-002-023		
04332+7102	TDS2891		1991	1991	1	136	136	2.8	2.8	11.79	12.46		+007-008	+007-008		
05182-6157	HJ 3755		1836	2009	5	286	274	20.0	21.0	8.3	12.8	K1II/III	+006+037		-62	456
07475-2942	HDS1101		1991	1991	1	106	106	0.1	0.1	8.82	9.58	K0III+...	-002+008		-29	4952
12186-1424	YSC 44		2007	2007	1	61	61	0.1	0.1	8.2	8.5	F2/3 III-	-054-005		-13	3514
21152+5351	BVD 134		1897	2006	11	58	57	23.0	22.8	10.51	11.67	K3IV- G5V	-014-060	-017-063		

FIGURE 2. Sample portion of the WDS summary catalog. See text for explanation of columns.

so we wish to get input from our "clients" as to the best changes to make and the best way to implement them.

A sample portion of the WDS summary catalog in shown in Figure 2. Columns here include the WDS designation (based on arcminute–precise epoch-2000 coordinates), discoverer designation, components (a blank indicates a simple AB binary), dates of first and last observation, number of observations, first and last value of position angle (θ), first and last values of separation (ρ), primary and secondary magnitudes, spectral classifications, proper motions of both components in mas/year, and a Durchmusterung (BD, CD, CPD) catalog name as a cross reference. (Not shown are additional columns for various note flags, as well as $0''.1$ precise epoch-2000 coordinates for each pair within a system).

Some of the limitations to the catalog are immediately apparent:

- **Components:** Discoveries of close pairs by high-resolution techniques have yielded some complex hierarchical multiple systems (for example, an "A" component resolved into close "Aa" and "Ab" components, then the "Aa" component further resolved into "Aa1" and "Aa2"). The five spaces now available are insufficient for component descriptors such as "Aa,Ab1" or "Aa1,Aa2". Addition of a few extra spaces easily solves this problem.

- **Separations:** Separations in the WDS now range from a few milliarcseconds to several degrees. Additional space and/or flags to indicate arcminute / milliarcsecond / microarcsecond separations will solve this problem as well.

- **Spectral types:** This field is insufficient for listing complicated spectral types for all components. The obvious solution again is more space. However, the WDS is not a spectral catalog, so given the amount of work which would be needed the addition of spectral information is of rather low priority.

- **Cross references:** The Durchmusterung catalogs have served as a useful cross

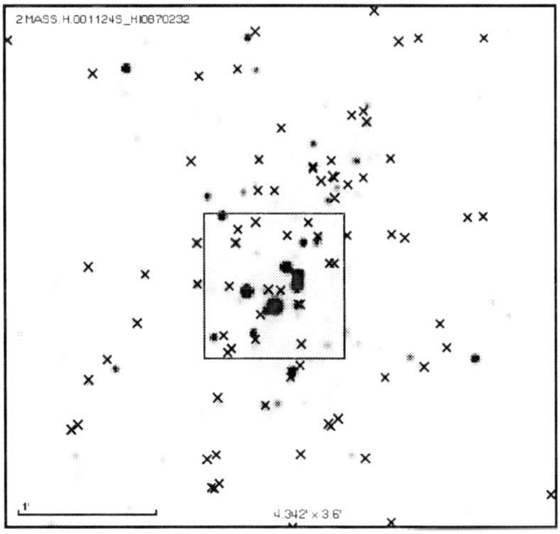

FIGURE 3. Aladin image of a small portion of the Orion Nebula, highlighting the limitations of the current WDS designation scheme. "X" symbols indicate the locations of binaries found during various duplicity surveys, while the square near the center of the field encloses a $1' \times 1'$ portion of the nebula. The current WDS designation is only able to accomodate a single system within each such square.

reference in astronomy for well over a century. However, many new pairs are fainter than the DM magnitude limits. Also, modern precise coordinates for many DM objects are not available. Possible successors may include:

- ⋆ Tycho? These are available for ~75–80% of objects in the catalog — a better choice than BD/CD/CPD, but probably not sufficient by itself.
- ⋆ UCAC4? This soon-to-be-released successor to the *Third USNO CCD Astrograph Catalog* will match perhaps 90–95% of the catalog. The UCAC catalogs are not well-known within the double star community, but should that be a factor?
- ⋆ NOMAD? The *Naval Observatory Merged Astrometric Database* incorporates the best available information from a number of catalogs (Hipparcos, Tycho, UCAC, 2MASS, USNO-B) and would match essentially all objects in the catalog. Again, it is little known within the double star community.
- ⋆ A mixture of various cross references? For example, HIP / TYC / 2MASS / DENIS / etc. could be listed, depending on magnitude.

Should we have cross references for one or both components of each pair? Additional identifiers for the secondaries of very wide pairs may be useful; the disadvantage is a larger, more unwieldy file.

Should cross references be deleted entirely and let the user rely solely on the precise coordinates of the primary of each pair? Alternatively, should a cross reference based on one or more of the above catalogs be removed to a supplemental file?

- **WDS designation:** The above mentioned limitations are essentially format issues, so simple (at least in principle) to address. This is perhaps the biggest problem, as is deals with the published designations for all visual binary and multiple systems. As noted above, the current scheme is based on arcminute-precise coordinates, so it is difficult to accommodate a density of more than one system per square arcminute. We can occasionally "fudge" the designation a little when two pairs happen to fall too close to one another. However, this doesn't help much in situations such as the one in the field of the Orion Nebula, as shown in Figure 3. Here we have as many as two dozen pairs within a square arcminute box. As the number of surveys for close, faint pairs increases, these situations will become more common.
  One obvious solution is to convert the WDS designation to one based on arcsecond-precise coordinates. For most of its history many WDS pairs were not known to that level (many epoch–1900 arcminute-precise coordinates were precessed to epoch–2000 without proper motion information; as a result many faint pairs were subsequently "lost"). Due to extensive work matching against other catalogs, often using Aladin, arcsecond or better coordinates are now known for 99% of all WDS pairs.
  This solution is not without its drawbacks, however. There is an enormous amount of information currently in publication; a new designation will undoubtedly lead to considerable confusion. (This is also the reason, by the way, why many component designations have not been corrected as further data have become available and additional components discovered.)

IMPROVEMENTS TO THE WDS DATABASE

The discussion above has been concerned with the WDS "summary lines" — the portion of the WDS database published on our website. However, the individual data lines also suffer from space limitations (largely the result of the 80-column limitations of the Hollerith punch cards from which the current database evolved!) Under their current format, there is insufficient space for separations and no space for formal errors in ρ, θ, or magnitudes, or for photometric filter information (other than codes for "blue", "red", or "infrared"). The solution now being considered is to expand the data lines and pattern them after those in the *Fourth Interferometric Catalog* ([3]).

EXPANDING THE WDS TO OTHER BINARY TYPES?

One question posed to the audience of this workshop was whether we should expand the WDS to include other types of binaries not currently cataloged:

- ⋆ Occultation pairs with only vector separations? (a code in the data and summary lines would indicate the vector separation and angle)
- ⋆ Astrometric pairs? (only summary lines would be added, with perhaps the value of the photocentric semi-major axis given in place of the separation)

⋆ Spectroscopic, eclipsing, and other pairs? (also listed only as summary lines with perhaps added columns for orbital period, etc.)

Audience members were unanimous in their opinion that such an expansion should not be undertaken, as it would make the catalog too complicated and could easily overwhelm those who maintain it. (This opinion was met with great relief by the catalog maintainers.)

TWO FINAL PLEAS

Recognizing the limited manpower available for maintaining the USNO double star catalogs, all observers are urged to email us their tables of double star astrometry, photometry, or orbital elements upon publication. This will ensure much more rapid incorporation of results into the appropriate database. Notification of any missing papers (by yourselves or others) is also welcome.

Also, observers are urged to include all necessary supplemental information — coordinates or other identifiers, dates, telescope aperture, filter information — with their published measures. Surprisingly, much of this crucial information is often missing from publications, sometimes rendering the measures themselves of little use to future researchers.

"The debt to our ancestors for the observations they made to our benefit, we can pay only by doing the same for our ancestors."

Ejnar Hertzsprung, 1961[1]

[1] Hertzsprung is the gentlemen dressed in white, standing next to Kaj Strand.

REFERENCES

1. Worley, C.E. & Douglass, G.G. 1997, A&AS 125, 523
2. Mason, B.D., Wycoff, G.L., Hartkopf, W.I., Douglass, G.G., & Worley, C.E. 2001, AJ 122, 3466
3. Hartkopf, W.I., Mason, B.D., & McAlister, H.A. 2001, AJ 122, 3408

CONCLUDING REMARKS

On behalf of the SOC, it is a great pleasure for me to make these remarks at the end of our mini-Workshop.

I would like to thank all of you for your participation and high-level contributions to the Workshop. Binary Stars is a vast field of research and we were able to hear presentations on different topics including observational data presented in the Invited Talks of H.Abt, N. Melikian, B.Mason, O.Malkov, W.Hartkopf, and Y.Balega.

I want to make special mention of the communication of the Pisco Group (represented by M.Scardia and J.L. Prieur). In this sense, I would stress the need for IAU Commission 26 to support their efforts. I hope the Officers of the Commission who are present at this meeting will share this opinion and that we can jointly support this group and others.

Also, I cannot forget the contributions of the OARMA team: the new research concerning exoplanets and exosatellites in binaries, the information in the OARMA Orbits Catalog, and the new MATLAB programming of Docobo's method for calculating visual double star orbits.

All members of the SOC and LOC deserve our gratitude for their effort to make this Workshop very successful. I hope to see all of you at the next meeting organized by our Commission.

Thank you very much!

J.A.DOCOBO
Chairman
President, IAU Comm.26

Workshop participants (from left to right):
First (lower) row: J.-L. Prieur, M. Scardia, J. A. Docobo, J. F. Ling, W. I. Hartkopf, N. D. Melikian, Yu. Yu. Balega.
Second row: H. Abt, V. S. Tamazian, M. Andrade, B. Mason
Third row: V. Lanchares, O. Malkov
Fourth row: J. R. González, M. Brea, P. P. Campo

List of participants

Abt, Helmut	Kitt Peak National Observatory, 950 N., Cherry Avenue, Tucson, Arizona 85719, USA
Andrade, Manuel	R. M. Aller Astronomical Observatory, University of Santiago de Compostela, Avda. Das Ciencias s.n., Santiago de Compostela, 15782, Spain
Balega, Yuri	Special Astrophysical Observatory, N. Arkhyz, Karachay-Cherkessian Republic, 369176, Russia
Brea, María	R. M. Aller Astronomical Observatory, University of Santiago de Compostela, Avda. das Ciencias s.n., Santiago de Compostela, 15782, Spain
Campo, Pedro Pablo	R. M. Aller Astronomical Observatory, University of Santiago de Compostela, Avda. das Ciencias s.n., Santiago de Compostela, 15782, Spain
Docobo, José Ángel	R. M. Aller Astronomical Observatory, University of Santiago de Compostela, Avda. das Ciencias s.n., Santiago de Compostela, 15782, Spain
González, José Ramón	R. M. Aller Astronomical Observatory, University of Santiago de Compostela, Avda. das Ciencias s.n., Santiago de Compostela, 15782, Spain
Hartkopf, William I.	United States Naval Observatory, 3450 Massachusetts Ave., NW, Washington, DC 20392-5420
Lanchares, Victor	University of La Rioja, Complejo Científico Tecnológico. Calle Madre de Dios, 51 26006 Logroño, Spain
Ling, Josefina F.	R. M. Aller Astronomical Observatory, University of Santiago de Compostela, Avda. das Ciencias s.n., Santiago de Compostela, 15782, Spain

Malkov, Oleg	**Institute of Astronomy of the Russ. Acad. Sci., 48 Pyatnitskaya St., Moscow 119017, Russia. Faculty of Physics, Moscow State University, Moscow 119992, Russia**
Mason, Brian	**United States Naval Observatory, 3450 Massachusetts Ave., NW, Washington, DC 20392-5420**
Melikian, Norayr	**V. A. Ambartsumian Byurakan Astrophysical Observatory, Aragatsotn province, Byurakan, 378433, Armenia**
Prieur, Jean-Louis	**IRAP CNRS – Université de Toulouse, 14 avenue E. Belin, 31400 Toulouse, France**
Scardia, Marco	**I.N.A.F. - Osservatorio Astronomico di Brera, Via E. Bianchi 46, 23807, Merate, Italy**
Tamazian, Vakhtang	**R. M. Aller Astronomical Observatory, University of Santiago de Compostela, Avda. das Ciencias s.n., Santiago de Compostela, 15782, Spain**

UNIVERSIDADE DE SANTIAGO DE COMPOSTELA

Observatorio Astronómico Ramón María Aller

International Workshop on Binaries

Binaries Inside and outside the Local Interstellar Bubble

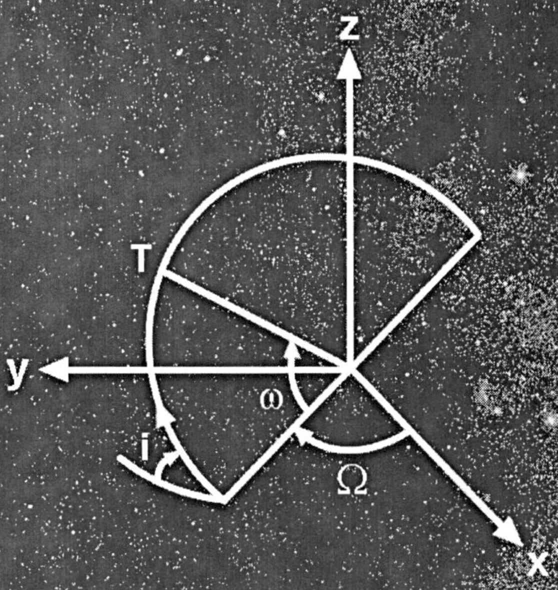

SCIENTIFIC ORGANIZING COMMITTEE

* *Yuri Yu. Balega*
  Special Astrophysical Observatory
* *José A. Docobo (Chairman)*
  Astronomical Observatory R.M. Aller
* *Brian D. Mason*
  US Naval Observatory
* *Vakhtang S. Tamazian*
  Astronomical Observatory R.M. Aller

LOCAL ORGANIZING COMMITTEE

* *Manuel Andrade*
* *Pedro P. Campo*
* *José A. Docobo (Chairman)*
* *Iván Fernández*
* *Josefina F. Ling*
* *Vakhtang S. Tamazian (Secretary)*

GOBIERNO DE ESPAÑA — MINISTERIO DE CIENCIA E INNOVACIÓN

Santiago de Compostela, Galicia, Spain
February 10-11, 2011

AUTHOR INDEX